About the Author

George has over 30 years' technology and business experience working in the telecommunications, financial services, and government sectors.

He is the founder and CEO of Wolfpack.com, a fintech self-directed trading, lending, and education platform. He took his previous technology company public in 2014.

In 2007 George was recognized by Bulletin Magazine as a member of their "Smart 100" list, which identified the 100 brightest individuals in Australia. He was listed in the same Top 10 technology cohort as the Atlassian co-founders.

He holds a Bachelor of Business (Computing), a Graduate Certificate of Management, and is a Graduate of the Australian Institute of Company Directors (GAICD).

George is married to Anastasia and has two children, Steven and Amanda.

HACKERS, CRACKERS, PIRATES AND PHREAKS

Disclaimer

This book is a memoir. It reflects the author's present recollections of experiences over time. Some names and characteristics have been changed, some events have been compressed, and some dialogue has been recreated.

Some parts have been embellished to varying degrees, for various purposes.

The content represents the personal views and opinions of the author and does not necessarily reflect the positions or opinions of any organization, institution, or individual with which the author is affiliated.

Neither the author, publisher nor any associated parties shall be held responsible for any consequences arising from the opinions or interpretations expressed within this book.

HACKERS, CRACKERS, PIRATES AND PHREAKS

Hackers, Crackers, Pirates and Phreaks

The Rise and Fall of the 1980's Global Computer Underground

A Memoir

George Parthimos

Copyright © 2024 George Parthimos.

First Edition. February 14, 2024.

All rights reserved. No part of this book may be reproduced or used in any manner without the prior written permission of the copyright owner, except for the use of brief quotations in a book review.

Paperback ISBN: 978 0 9756119 0 6

Ebook ISBN: 978 0 9756119 1 3

For Mum and Dad.

HACKERS, CRACKERS, PIRATES AND PHREAKS

Contents

Houston, We Have a Problem 1
Ground Zero 7
My Background 11
Computers are Stupid 15
Snake Byte 27
Level Up 37
On Your Bike 59
Luck's a Fortune 65
A New Online World 75
Going Global 93
How to Hack and Crack - Retro Style 125
Pecking Order 143
Myths and Legends 159
A Pirate's Life 163
History Repeats 175
Citibank Hack 181
Life After Citibank 197
You're Hired 205
More Than a Game 215

HACKERS, CRACKERS, PIRATES AND PHREAKS

Houston, We Have a Problem

A message splashed across the screen: "We need to talk!".

"Damn!", I thought. Not now.

I had just spent hours downloading the latest computer games from a Bulletin Board System in Germany, and was about to start uploading onto Zen, Australia's premiere computer underground site.

The copies were freshly minted by a software cracking team during their latest weekend catchup. This time, they all travelled to the Netherlands, a short drive for the Germans, Austrians, and Belgians.

A rogue team so skilled in their exploits, they were known as the best software crackers in the world.

They would often invite me to attend, but travelling from Melbourne to Amsterdam was a little tricky for a

17-year-old high school student. Plus, I didn't have a driver's license.

I had to kick off my uploads onto Zen before someone else beat me to the punch. Being the first to upload newly pirated games gave you incredible street cred in the community. Literally within 48 hours of a new game's release, you had possession of a commodity that everybody wanted. The games were so hot, they weren't even scheduled for release in Australia for several more weeks.

"Rebel, come to Chat 109 now!".

Wow, this guy was persistent!

It was a typical warm 1989 summer's evening in Melbourne Australia. Most 17-year-olds were out with friends, sneaking into bars and clubs with fake IDs or hanging around the local street drag racing hotspots near Bouverie Street Carlton, a mecca for the euro slicksters of the day.

My mates had called home several times but couldn't get through. The phone line was tied up for hours.

No time for chit-chat, important work was afoot.

A few keystrokes later, and the process had begun. New games were being uploaded which kicked off national distribution to whoever wanted them. Completely free of charge.

"Rebel, come now! Chat 109!".

"I'd better get to Chat 109 and see what all the fuss was about.", I thought.

As I entered the private chat, I greeted the impatient hacker who had been harassing me for an urgent chat.

"What's up…" I began to type. Before I could even finish my sentence, I received a barrage of messages.

"Stop whatever you're doing and get off Zen NOW!"

Hmmm…. Wonder what this is about.

"We need to meet tomorrow at the usual place. 4:00pm. Make sure you're there."

Then, just like that, he was gone. Unplugged. Disappeared as quickly as he had arrived.

I had no phone number, no way to contact him other than in the usual way. So, I decided to finish the uploads and hang out in the online lounge until some of the other pirates or hackers turned up for their nightly visits. Everything seemed normal enough. Zen was humming along like a well-oiled machine.

Except, something was odd. The usual suspects weren't coming online. Strange. This was rush-hour for Zen.

The next day I caught the train to Melbourne's central business district and arrived at 4:00pm at the usual meeting point: The City Square, corner of Swanston Street and Collins Street, Melbourne.

There he was, nervously waiting for me. He wasn't alone. Others were there from the community. And they looked just as nervous.

Walking up to the small group of fellow teenagers, it was clear there was a problem. What came next completely blew me away.

"Rebel, we're fucked! Zen and Pacific Island have been infiltrated by the FBI and the Australian Federal Police. They've been monitoring all our activities for at least three months and are about to start arresting people."

Wait… WHAT?!?!

"Remember the Citibank hack last year, well, they've traced it back here. To 'The Realm'. And they're going after everyone connected with them. They have all the system logs. They know everything. They've gotten to Craig, and he's been helping them for months!"

Craig Bowen was the owner and system operator of both Zen and Pacific Island Bulletin Board Systems, the two biggest computer file sharing and chat sites in Australia. Both systems were run from his bedroom, and they were situated next to each other.

"Have you chatted to anyone called Stuart Gill?", he asked, short of breath and sounding more desperate.

I hesitated. Stuart Gill was a new online name which had turned up a few months earlier. He couldn't have entered the private sections of the bulletin boards without a personal invitation from Craig. This was an exclusive club reserved for only the cream of the crop. The best and brightest hackers and pirates from around the world.

The name was suspect to me from the beginning. Everyone on the boards had pseudonyms. Rebel, Masked Avenger, Interceptor, Electra, Killer Tomato, Ivan Trotsky. No-one ever used a real-sounding name.

"Yeah. He has been pestering me for months in the chats. I've been avoiding him. No time. Too busy!' I responded, deep down knowing what was coming next. My pulse quickened.

"He's a Fed! He's been gathering information for the Australian Federal Police under a joint taskforce with the FBI."

"Holy shit!", I thought. This is not good. Our fingerprints were all over the Zen and Pacific Island exclusive hacker and pirate areas.

"What the fuck do we do now?"

Ground Zero

We are all very familiar with the Internet today, the benefits it offers to our everyday lives, and the continuous innovations and developments being accomplished. The Internet as a commercial platform came out in the early 1990's, and experienced significant market growth and consumer adoption in the late 1990's during the 'Dot Com Boom'.

The history of the Internet dates back to the US Department of Defense (or 'DoD') in the 1960's, who developed the network protocols and many of the software and services that would later be released into the public domain. These early innovations would form part of the actual foundation of the Internet.

The story of the Internet is well-known, well-documented, and an undisputed historical event. The two stages which defined a global connected computer system were the DoD innovations of the 1960's and the consumer mass-adoption of the 1990's.

However, few people are aware of what happened between these two stages: the transition from mainframe corporate computers to home users, and the evolution of the standalone home computer to a connected system.

This is a very narrow period which helped shape the Internet into what it is today. It was critical in the evolution of the Internet. This period in the 1980's ushered in the dawn of the Bulletin Board System (or 'BBS').

These BBSs were home computers that had dial-in phone lines, allowing any civilian to access a connected computer environment, which had previously never been available to the general public.

It was revolutionary for its time, and allowed computer hobbyists and enthusiasts to build the foundations for what later became the Internet.

If you ever wondered how the Internet evolved from a restricted defense or corporate computer network to a public information superhighway, this is genesis. Ground Zero.

The people in this book were involved at the coalface. Innovators and early adopters of a technological revolution which has shaped our connected world.

During this period, a global computer underground emerged which would push the limits of software pirating, computer hacking, and telephony phreaking. Its exploits would become legendary, and the repercussions of its actions would result in the formation of the laws and technical measures designed to protect the public from the activities of these trailblazers.

As with any revolution, the seeds are planted with much excitement and anticipation of what could be. Built on the back of foot soldiers whose initial intentions were good, but later corrupted by the power of information and privilege. It's a tale as old as time.

If you were using pirated software in the 1980's, there was a very strong chance that the source of the software came through this tight-knit underground community.

Being involved in the life was addictive. It was exciting. Much more exciting than the mundane teenage life I was living. I found freedom, respect, a sense of belonging, and an intellectual high very few would ever understand or experience.

The online world became my obsession. In the process, I lost who I was. I lost touch with reality. The more I connected to the online world, the more disconnected I became from the real world around me.

This story covers a six-year period between 1983 and 1989. I have never shared this story in detail with anyone.

However, 40 years have passed. It's time to tell my story.

These are my memories and experiences of the global computer underground, and what happened in the 1980's that led to the dawn of the Internet.

My Background

I was born into a Greek immigrant family. Both my parents were from small villages near Sparta, each having a population of about a thousand people. They immigrated to Australia in the 1960's, arriving with the typical migrant story: a suitcase, a little bit of money in their pocket, basic language skills, and not much else.

Most will know Sparta from the popular movies depicting Leonidas and his brave 300 soldiers who took on the might of the Persian Empire in a 'David versus Goliath' battle. It's a story of extreme sacrifice, honor, bravery, loyalty, unity, and respect.

Born as a middle child, one of three brothers, we grew up in the working-class suburb of Thomastown in the north of Melbourne. In the 1980's, Thomastown was a rough neighborhood. There were several street gangs floating around including the 'Street Runners' and 'Thomo Sharps'. The area bordered other rough neighborhoods such as Broadmeadows and Lalor, which had their own fair share of crime.

Growing up in Thomastown, you learned pretty quickly how to survive. The trick was to network with the local kids in the neighborhood. These kids would either be members of the gangs themselves or have brothers or cousins involved. By putting yourself out there, you got to know the people in the community, and more importantly, they got to know you.

Australia in the 1970's and 1980's was a very different place from today. As kids, we used to go outside and roam the streets, playgrounds and parks looking to socialize with other kids in the neighborhood. The golden rule from my folks was simple: When the sun goes down, start heading home. The rest of the day was yours to explore and do whatever you liked.

We would all meet at a local place around the same time. Most cases, it was the eastern agora of Thomastown High School. Typically, there would be 20 or 30 kids ranging from 10-year-olds (usually younger brothers of kids who had strict instructions from their mum to look after their younger siblings) up to 18-year-olds. When enough people turned up, a game would usually break out. Cricket during the summer, Australian Rules Football during the winter, and Basketball or Soccer every other time.

Through socializing and networking, you got to meet a lot of people, and hear a lot of chatter about the streets. Rumors of pre-arranged gang fights, car break-ins, stolen goods for sale, you name it. A typical day hanging out with the street kids of 'Thomo'.

At Thomastown High School, I was a strong student academically; straight-A's, particularly in English, Math, and Science. I excelled ahead of other students in my year level, to the point where the school offered to put me into Year 11 Math, English, and Science. For some reason, I was ahead of everybody else.

School came easy to me. I had the ability to remember things, particularly specific details, which many didn't. I don't have an eidetic memory by any stretch, just the ability to remember things and regurgitate them quickly and succinctly.

In fact, I was offered the opportunity to participate in a scholarship program with an exclusive private school in the northern suburbs called Ivanhoe Grammar. At the time, Ivanhoe Grammar had a program where they would assist underprivileged kids who had a strong academic level, to attend school on a full scholarship.

Thomastown High School was seen as one of the more socially challenged schools in the state, so it was a potential feeder school to the scholarship program.

When I went to Ivanhoe Grammar to have a look at the school, I quickly realized that a private school wasn't for me. I didn't feel I would fit in.

The students attending the school came from very affluent backgrounds. Their parents had high disposable incomes, large luxury homes in desirable suburbs, and top-of-the-range prestige vehicles. They would holiday overseas, go skiing in the winter, wear designer clothes, and take music lessons in between their tutoring.

We were very much a working-class family. We didn't have the luxuries of the Ivanhoe Grammar cohort. Our car was a beat up 15-year-old Australian made 'classic', and we lived in a working-class suburb. We certainly weren't at that exclusive private school level.

I chose to attend St. John's Greek Orthodox College for my final four years where I graduated High School. After graduation, I went to university where I graduated with a Bachelor of Business majoring in Computing.

Computers are Stupid

In 1983, as a fresh-faced Year 7 student at Thomastown High School, I was introduced to my first Computing class.

We had all heard about computers, mostly from Hollywood movies, and believed they were highly intelligent machines. We also knew that computers were something we all needed to learn about. It was, after all, the future.

Thomastown High School was rated one of the worst schools in the state of Victoria. The school catered for the growing European migrant families which were drawn to the outer northern suburbs of Melbourne looking for affordable housing within close proximity to their work. Most were factory workers, typically doing 10-hour shifts up to six days a week. Working class families who had answered the call to come to Australia to help fill the growing labor shortage of the 1960's.

The area was expanding so quickly that the school had been hastily constructed by the Education Department using the same blueprint as many other schools in surrounding suburbs. Grey concrete brick block work with a 'Mission Brown' colored corrugated iron façade and roofing, they were built more for function than form.

The school had a horizontal figure-eight design. The administrative offices separated each of the rectangular concrete agoras which were encircled by single-story buildings. The computer lab was located next to the administration offices; close proximity to the head honchos so they could ensure nothing went missing on their watch.

Computers were expensive assets, and the school had benefited from its low sociodemographic status with grants to buy ten Apple II computers. Someone somewhere in the Education Department figured it might be a good distraction from the gangs and street crimes which had engulfed the area.

Each classroom was identical in design and layout, probably to keep the costs down during construction. Large floor-to-ceiling dark brown timber windows lined the classroom perimeters, allowing ample natural light to flood in. It also ensured hot summer days were amplified by the glasshouse-like design characteristics.

Thankfully, Melbourne weather was cool most times of the year, so the sweltering heat was relatively infrequent.

The teacher, in front of some 30 students, had set up an Apple II computer at the front of the classroom. We all looked at this computer with amazement and excitement. It had a screen, a keyboard, and a floppy disk drive.

Our adolescent minds raced with visions of talking machines controlling spaceships and launching rockets to Mars.

"They must be the most powerful things on the planet!", we thought.

Hollywood had really set high expectations on what we were about to experience!

The teacher then asked a simple question: "Who thinks that computers are smart?"

Dozens of hands eagerly flew up in unison. What else would you expect from a room of 12-year-olds with preconceptions based on the blockbuster movies of the day.

"Yes, computers are smart", we all thought.

The teacher then typed a few words on the keyboard and revealed the monitor display: "I am stupid". He then asked "Well, if computers are so smart, why are they letting me write this?"

That very simple example showed me that, essentially, computers do whatever we tell them to do. Computers are a tool that we control. They were not smarter than us. We make them what they are.

That realization for me, as a 12-year-old, was very powerful.

During the late 1970's and early 1980's, access to anything digital was very limited. We had basic electronic gaming systems such as the Atari 2600, a game console with a cartridge slot that allowed you to insert and play games such as Pitfall, Breakout, and Pong. These were very basic but highly addictive games. We also had access to hand-held games which were making their way from Japan. Donkey Kong was a favorite amongst the kids of the day.

Computers in the home were extremely rare, mostly because they used to cost a fortune and needed expensive software to be useful. They were designed for business use. The idea of having a computer in every home seemed far-fetched, although people were talking about it as a concept.

Seeing a computer in real life was exciting. I remember thinking to myself "Wow, this is so cool!".

I proudly proclaimed to my parents that I wanted to get involved in computers. I explained to them that this was going to be my career path.

So why did I pick computing as my desired vocation so quickly? Well, it's very simple.

Coming from a European migrant family, we learned to do everything ourselves. We were plumbers, electricians, spray painters, bodywork professionals, mechanics. I cannot tell you how many times I would be in the garage doing something with my dad. It seemed like a nightly ritual!

Well, after tiling and plumbing and spray painting for what seemed like an eternity, I quickly realized that whilst I could do it, I hated it. I hated getting grease under my fingernails, getting spider webs on my head, getting oil all over my hands, and getting glue stuck to my arms. Not to mention the constant grazing of knuckles while trying to loosen a stubborn bolt.

Whenever I complained (and I complained very often), my dad would say "If you don't want to do manual work, then make sure you educate yourself so you can work in a nice clean office". That always stuck in the back of my mind.

I often wondered whether dad was trying to teach me a lesson. He knew how much I hated the work, but he persisted in getting me to do things. A part of me thinks this was his way of making sure that either I educated myself so I wouldn't have to do that type of vocation, or I learned from an early age how to work with my hands.

He would often tell me "A man who knows how to work with his hands will never go hungry".

As soon as I saw a computer say it was stupid, I was sold. That was my ticket out of the working-class life.

I convinced my parents that this was the path I wanted to take. Of course, they were excited because I had found an academic path I wanted to pursue, and they could help me achieve my objective by buying me a home computer.

We went to the local Kmart, and dad bought me my first computer: A Commodore Vic 20. We purchased it during Christmas of 1983. It cost $220, which was almost a week's salary at the time.

The Vic 20 was a rectangular-shaped beige unit with a coffee-brown built-in keyboard. It had a meagre 3.5 kilobytes of RAM (or 'Random Access Memory', used to store programs), and 1.5 kilobytes of ROM (or 'Read Only Memory', used for the operating system). It included an external tape drive that allowed me to store any programs I created or copied.

I took the computer home, set it all up, and was instantly hooked.

At the time, I shared a bedroom with my younger brother Con. Peter, my older brother, got his own room because he was the oldest and therefore had certain privileges ahead of the rest. He was in First Class while we were squeezed in Coach.

Our shared room was cozy. It measured 11 feet by 10 feet, with 8-foot ceilings, and had two single beds on either side of the room. In the center against the wall sat our desk which we used for schoolwork. It was here that I set up my computer, much to Con's objections. He wasn't complaining that his workspace had been seized, more that he had lost valuable shared territory in the uneasy truce we both had.

The room had been carved up into zones. Con had his side and I had mine. These areas were restricted; one couldn't cross the border to enter the other. Down the center was the neutral zone, where we both had agreed to use to access the door to exit the room. The small built-in closet on the other corner was also neutral.

At one stage, we stuck tape on the floor to clearly define our territories, but mum ripped them up one weekend when she couldn't vacuum the carpet.

The desk therefore was valuable real estate. I needed to barter with Con for him to agree to alter the treaty. My currency: games. I allowed Con to use my computer whenever I wasn't using it to play whatever he wanted. I only had one rule: no food or drinks while using the computer. He agreed, and warring tensions eased.

Inadvertently, Con would become my games tester. Every time I would get the latest batch, I would get him to play them and let me know what he thought. He became obsessed with playing games, so much so that he would often stay up after midnight playing.

When you're in a small dark room trying to sleep, all you see is the flicker of the bright screen and the sound of a crunching joystick. Sleep often suffered.

Deep down, I actually enjoyed sharing a room with my younger brother. We were only 13 months apart, but we were very different people. I was a perfectionist and would insist on folding my clothes neatly on the chair or in the closet. Con, on the other hand, was more liberal. His clothes would be sprawled across the floor wherever they landed. Hence why the borders were drawn up. My side was always neat and tidy, Con's was always a mess. I think he did that just to piss me off, as younger brothers tend to do when they want to stick it to their oppressor.

After setting everything up on our desk, the first thing I did was start writing code. I bought computer magazines and absorbed everything about the Commodore Vic 20. I was particularly interested to learn how the memory and the tape storage device worked. I basically researched everything I could find about my new home computer.

Unbeknownst to me, the Vic 20 was approaching the end of its production life, as a new improved computer was being launched to replace it the following year. But this didn't faze me; I had a computer in my possession which I could learn from. And I certainly wasn't going to tell my dad he had just shelled out most of his hard-earned weekly salary on a dud!

At the time, the only way to gain access to anything related to the Vic 20 was through magazines. I had tried the local library to see if they had anything on their shelves, but it was a bust. I did, however, manage to find some BASIC computer programming language books which came in handy.

Scouring through the newsstands, I would grab any magazines which had Commodore Vic 20 source code printed in them, and physically type in every single line of code onto my computer.

Software and games were expensive, and so I had to improvise. A computer magazine would cost a few dollars, whereas a game would cost $40 or $50; Out of reach for a 12-year-old son of blue-collar migrant workers.

The source code was usually 10 or 15 magazine pages long. I would sit there for hours manually entering the code in my computer. And of course, I would make typing mistakes which needed correcting. I would debug the code to make sure it worked as per the magazine screen shots.

Once it was 100% accurate, I would start tinkering with it. I would make additions and changes such as the colors, layout, and logic.

After I had completed my changes, I had to save the code. The tape drive which accompanied my Commodore Vic 20 was the only way I had to store what I had created. The storage method was effectively a regular 'compact' cassette tape that was inserted into the external cassette deck.

I needed to line up the tape drive counter, press reset to "000", press the RECORD button, and the program would be stored on the actual cassette tape. Next, I had to write down the counter start and stop number so that I knew where exactly on the tape the code was stored so I could recover it later.

For example, if the counter was set to "000" and then it went to "013", these 13 counts were where the code was stored. I then allowed 2-3 counters buffer before storing the next piece of code. From "016" to "025" and so on.

There was no catalog track or directory. It was a pencil and paper with the counter number ranges and the code title/information.

Whilst very crude, it was mind blowing!

HACKERS, CRACKERS, PIRATES AND PHREAKS

Snake Byte

Years 7 and 8 were my computing foundation years.

Thomastown High School had a computer lab, as most schools did at the time. The lab contained ten Apple II computers. During lunch breaks, students would either study in the library, play sports, or socialize in the cafeteria. Some, like me, would hang out at the computer lab.

The lab was located in the center of the administration building, in the heart of the school. The all-white painted room measured 20 feet by 20 feet and had two entrance doors at either end. 'Standard Issue' dark brown imitation woodgrain desks lined the two adjacent walls, each holding an Apple II computer, monitor, and a floppy disk drive. Each desk was assigned two chairs, allowing up to 20 students to be squeezed into the tight space.

At one end of the room, the Computer Lab Assistant sat on a separate desk overseeing the computers. A clipboard containing a pencil and the booking schedule was located next to him.

He was the gatekeeper of the lab, and he knew his importance in the pecking order. A geeky student by day, his status elevated greatly when he took his seat at the head of the lab. Any bookings or disputes went through him. Like King Solomon's Throne Room, he was judge, jury and executioner.

The room had no windows, which meant no natural light. The only light source was a row of fluorescent bulbs down the middle, which poorly illuminated the room. Some occasionally flickered indiscriminately, others had gone to meet their maker.

Ventilation was so poor, both doors needed to stay permanently open to try and get a draft flowing through the room. This obviously didn't work too well; the combined heat from the computers and the students gave the room a distinctive aroma. A cross between a sweaty gym and a car engine bay.

Given its layout and close proximity to the school administrators, it was probably once a storeroom for student records, filing cabinets, and a graveyard for broken tables and chairs.

It wasn't comfortable, but it was bearable. A dreary looking place, it was more dungeon than classroom. But we were drawn to its hidden treasures.

Accessing computers was very popular back in 1983, and most schools had implemented strict usage policies for students. You had to pre-book your allotted time slot and desired computer (they were each numbered, so you had to pick one and sign up to use it). The Computer Lab Assistant would monitor usage and ensure people jumped off when their time was up. The system was designed to allow as many students as possible to use these incredible machines.

And what would every student want to do when they had their time on a computer? Play games of course!

The main game at the time was called 'Snake Byte'; a game of skill where you used your up, down, left, and right keys to move a snake around a screen and eat certain items like fruit and bugs. The more you ate, the larger the snake grew, making it more difficult to maneuver around the screen without hitting a part of your body. Once you ate all the available items, an exit door would appear which you had to successfully get through without hitting yourself or the walls.

It was a challenging game which required tremendous skill, patience, and hand-eye coordination (the three things I lacked). Everybody in the lab was hooked on this game.

When you walked into the lab, the first thing you needed to figure out was how to get involved in the scene. Most students playing the games were focused on their screens, their eyes wide open, absorbing every move as the reflection of the screen danced across their face. Conversations were limited to a couple of congratulatory comments when someone got through a stage in the game, or banging of fists on tables as their snakes crashed into a wall.

Sometimes general banter would break out when someone achieved the high score for the day. Like a town crier, they would proudly announce their score to the room, and then stretch out their back and neck muscles before cracking their knuckles to welcome the next challenge.

I walked in, naively asking "Oh wow, these games are cool, can I play?". Without taking their eyes off the screen, most would say "No, I'm playing it!" or "No, I won't let you use my disk!".

You quickly realized that if you didn't have the basics, you were an outsider.

The only way you could access a computer and play games was if you had your own 5-and-a-quarter-inch floppy disk to store the game on. Only then could you go and play the game when you wanted.

But of course, trying to barter for the game when you didn't have a floppy disk was pointless. People would not give you the time of day because effectively you were not of any value to them. After all, you were just some dumb kid hanging around the lab wanting to play a game. There were plenty of those kids out there.

If I was to join this community, I would have to purchase my own floppy disk and get a copy of the game 'Snake Byte'. If I could do this, then not only would I be able to do what I wanted, but others would accept me within the community. Having a game that people wanted to play and a floppy disk were the keys to success.

Unfortunately, these disks cost $10 a-piece at the time, which was a lot of money.

It's at this point that I started to understand that there was a certain pecking order when it came to computers. Whilst you might have an interest in it, if you didn't have the basic tools, you were not going to get too far.

Armed with this basic understanding, my first task was to go and buy a floppy disk. Thankfully the school sold these at the administration office.

Next, I needed to save up $10 bucks. At the time I was getting a $2 daily allowance for my lunch money, so I set about saving up a little bit every day. Lunch wasn't as important as getting my hands on this elusive disk. I reduced my daily expenses to $1, which allowed me the luxury of buying a small serving of fries and a drink. The other $1 went into the piggy bank.

After two weeks of saving, I managed to scrape together the $10 bucks I needed and bought my first floppy disk.

That was half the journey. The next important part was to get a copy of the game.

I began approaching people in the computer lab who had a copy of 'Snake Byte' and tried to barter with them to get a copy. I quickly realized that most of them were just gamers and had no idea how to copy a game. After some prodding, I managed to get the names of the people in the room who had the game and the ability to copy it. These were the suppliers, and they were the most revered people in the lab.

To copy a piece of software, you needed to have two things. First, you needed the game you wanted to copy, and second, you needed a program which copies the game.

This was a new concept to me. I had always thought that you just stuck a disk in a drive and then, magically, the game would be transferred from the computer to you.

I finally found the right person in the lab who had both the game and the means to copy it. But it wasn't free. I had to fork out $1 for the game.

It seems the international language of cash works well even in a high school computer lab.

But simply getting a copy of the game wasn't enough. I would effectively be the same as the other gamers in the room, and that's not where I wanted to be. I wanted to be the guy people came to so that they could get the games.

So, for an extra $1, I also got a copy of the program used to copy the games.

"Lunch today will have to wait", I thought. This was far more important to me.

Next thing I knew, I had managed to get a copy of the game and the copying software on my disk.

Success!!

I was now at an equal level to the top guys in the room. Every time I walked into the room, I had my floppy disk, I had my own game, I had the copying software, and I could speak at the same level as everybody else.

Having a disk and a game represented power and, more importantly respect, in the lab.

Suddenly I started to understand the basic dynamics of how the world works.

From this very simple lab transaction, I quickly realized that information is power. The more you know, and the more you have, the more desirable you are, and the more you can barter and negotiate with people.

It's a simple lesson of the 'haves' versus the 'have-nots'. Those who possess something desirable are treated very differently to those who don't have anything valuable. Remember, value is not necessarily monetary (although in most cases it comes down to money in the real world). Value can be perceived in many forms. If you have something that someone else desires, you are important to that person. You can then control the situation to gain a strategic advantage.

It was a very important lesson to learn, which would lay the foundation for the rest of my teenage years and moving into adulthood.

HACKERS, CRACKERS, PIRATES AND PHREAKS

Level Up

By 1985, my Commodore Vic 20 was pretty much obsolete. Commodore had released their next generation computer called the Commodore 64, which was about 30 times more powerful than my Vic 20.

Although I learned a lot with the Vic 20, it was time to upgrade.

After succeeding in my quest to level up in the computer lab, I noticed that everybody was obsessed with using the Apple II. It was the most desirable computer in high schools. So, I decided it was time to upgrade to an Apple II for home.

It was time to level up again.

At the time, I was at the same level as the top guys in the lab. The next level up was to either have other games that people didn't have, or to have a computer at home so you weren't restricted by 20-minute usage slots allocated in school labs. I, of course, wanted to have both.

The first thing I did was start lobbying my parents to upgrade to an Apple II computer. Of course, we couldn't afford an original Apple II computer because it cost roughly $2,000 at the time which was very expensive.

However, we could afford an Apple II clone. This was effectively a knock-off Apple II computer built in China. It had the same look and feel as the original, with some quirky differences.

For starters, the clone had a separate daughterboard which plugged into the main motherboard. This daughterboard contained the CPU and BIOS. An original Apple II computer had all the components soldered on the main board. It also had some very slight variances in the chipset components. They weren't all original components, so therefore sometimes misbehaved when you used certain software or commands.

I used to hang out at Rod Irving Electronics on High Street in Northcote, where I would buy a lot of my electronics parts that I needed to build various projects.

For me, walking into Rod Irving Electronics was like walking into a toy store. The single-fronted shop had a rectangular shape which measured 20 feet by 120 feet. As you walked in, you were greeted by a large, long glass counter which ran along the left side of the shop. It housed many of the more expensive electronics items such as disk drives, memory chips, multimeters, soldering irons, and DIY kits. Next to the counter sat a large bookshelf containing the latest computer magazines, electronics books, and technical reference manuals.

A row of timber shelving lined the opposite wall, displaying various electronics components such as resisters, ribbon cables, LED lights, twisted-pair copper cables, and chip sockets. The rear of the shop showcased a large range of shiny new home computers, all quietly humming away and ready to test drive.

Soft elevator music echoed through the shop as you browsed the shelves and cabinets.

My friend Steve and I would sometimes skip school and head down to the store, which happened to be around the corner from his house. We would play with the new computers on display and talk to the store attendants about the latest electronic kits available for sale.

I remember once building a Morse code system from some old shoe boxes. I ran a twisted pair copper cable to my brother Peter's room next door. We would send messages to each other, even though Peter didn't know Morse Code. For me, it was a huge thrill to build something from various bits and pieces and get it to work.

Rod Irving Electronics had an Apple II compatible computer, and so I continued lobbying my parents in early 1985 to buy it for me. It wasn't easy. Although the computer was half the price of the original, it was still around $1,000 which was almost two weeks salary for my parents at the time.

This was a big sacrifice for our family. It meant that we would go without other things. Important things. My brothers would bear the brunt of not getting things they wanted. I knew it would affect everyone, but I was slowly becoming obsessed with computers and making sure I had the best possible technology at my fingertips. Yes, I could have settled for a much cheaper Commodore 64 which was half the price, but this to me was a toy. It wasn't in the same league as the Apple II.

Thankfully I managed to convince my parents it was the right thing to do to help me advance with building my computer knowledge. They could see I was passionate about computers. I had practically worn out the Commodore Vic 20 keyboard with all the miles I had done on it.

My younger brother Con wasn't too thrilled with me getting all these cool things, but he quickly fell into line when I allowed him to play games on it. After all, he had access to a games console in his bedroom.

Suddenly, he became quite popular with the locals!

Games didn't matter to me. I was only interested in one thing: Levelling up.

My Apple II clone computer had one floppy disk drive, 64 kilobytes RAM, and a monochrome monitor. Very crude by today's standards, but back in the day I was thrilled!

Having this level of technology at home allowed me to start coding and playing around with stuff in my own time. No more 20-minute egg timers, dirty grubby lab keyboards, hot smelly lab rooms, or Computer Lab Assistants telling me to get off. This was a big step up.

I now had the same state-of-the-art home computing power as all the other computer enthusiasts.

Places like Rod Irving Electronics had all the bits and pieces you needed if you were into computers. And the number one item on my wish list was a 10-pack of five-and-a-quarter-inch floppy disks.

There were two brands: Xerox and Verbatim. I personally preferred the Verbatim brand because it had a purple/navy blue color and a cool logo design. It was also $10 cheaper per 10-pack than the Xerox brand.

Verbatim disks were selling for around $60 for a 10-pack, or roughly $6 per disk. Again, saving my lunch money and doing some chores around the house, after a couple of months I managed to scrape together enough money to buy my first 10-pack of Verbatim floppy disks.

Suddenly I had a major dilemma. I had all these expensive blank disks, but only one game.

I decided to level up again. I needed to source games from other people outside of my existing network.

If I could source other unique games, I would shift from a consumer to a supplier. I would become the person who actually provided the games people wanted.

It was the next logical way to level up.

If I could achieve this, I would become the most popular kid in the area.

Armed with a box of 10 blank floppy disks and one game, I started looking around the computer lab to see what else was available. I realized that whilst 'Snake Byte' was the most popular game in the lab, there were a few other games being played.

I managed to become friends with a few people, which led to me getting copies of these games. I'd gone from one game to four or five. Slim pickings for what I was after.

Through discussions with my cousins at various family functions we attended, as well as friends who attended other schools, I realized they didn't have the games we had in our lab. Importantly, other games were being played in their school computer labs.

That was really interesting to me. My world had evolved around one computer lab at one school, which had a limited number of games. I now realized there were other labs at other schools, as well as other places such as public libraries, which had completely different games.

Using my finely tuned networking skills developed on the mean streets of Thomastown, I managed to get introductions to people at Northcote Library, which happened to have Apple II computers in its study area. It turned out that Northcote Library's game options were very different to the options available at Thomastown High. They didn't have the games I had, and I didn't have the games they had. It was the perfect environment to swap games with people from Northcote.

It then dawned on me: if that's the case between two computer labs in the northern suburbs of Melbourne, it's probably the same case in other parts of Melbourne. I realized that there was a way to get access to a much wider array of games. It was a way for me to expand my reach and grow my collection.

And so, I started attending computer meets that operated monthly, which were recommended to me by my new network. It turned out that these computer meets were attended by a wide cross-section of the Melbourne computer enthusiast community.

Meets were normally held in local community centers, places where the elderly would attend ballroom dancing events and local women would turn up to weave scarves and socks for their grandkids. White walls, grey carpet, open spaces. They were pretty dull and dreary places, almost like a forgotten closet out back which housed your old gym equipment.

A familiar detergent smell would lightly linger throughout the room. "Local council cleaners must have come through for a quick tidy-up between events", I often thought.

Walking in, you would pay an entry fee to help cover the small overheads, usually a buck. The friendly organizers at the front fold-out beige plastic table would greet you with a smile, and hand over a ticket stub in exchange for the entry fee. Inside, more beige fold-out tables lined the perimeter of the room. Usually fifteen to twenty vendors displaying their various items for sale: Books, magazines, reference manuals, and of course, original games.

It was a strict cash-only policy.

As you surveyed the room, you realized most of the vendors and patrons were regulars like you. Everyone was on a first name basis. Some vendors had set up their own computers on a separate fold-up plastic table, ready to demonstrate the games they were selling. Patrons would bring along their games collections and floppy disks, fully expecting they would come in handy during the event.

You knew which of the vendors were amenable to swapping games. One look, a wink and a nod, and it was on.

More games and other software applications would pop up at these events which I didn't have. I would build out my growing catalog of software by trading with what vendors had to offer. In addition, they were happy to allow you to use their systems to swap games with other patrons. After all, we were all there for the same reasons.

Swapping games was a tingling sensation to me. Although these were copies, the excitement of getting something shiny and new filled my heart, strangely similar to opening presents as a child on Christmas morning. I could almost feel the fresh cellophane wrapper peeling off the bright colorful packaging design of an original game, releasing a unique aroma of manufactured plastic and paper. It was intoxicating.

Without realizing, I had become a pirate, which was the term used to describe individuals who traded games in large quantities.

My reasons for becoming a pirate were innocent enough. Firstly, I enjoyed computers. Secondly, my working-class family background couldn't afford to purchase the software I wanted. And thirdly, I got to meet a lot of people in the industry. I listened and learned from these people from across different parts of Melbourne.

Remember, computers started out as my hobby. They were now quickly becoming my obsession.

As I built up a vast catalog of software titles, I saw myself as a collector. Like a wine connoisseur buying rare bottles to store in a cellar. Not for consumption, just to admire.

I didn't consider myself a pirate, even though I was illegally copying and distributing software.

Before I knew it, I had a treasure trove of 70 or 80 disks containing hundreds of games and software applications. Overnight, I became the most popular kid in school, purely because I had access to all the latest games and software which no one else had.

People would call me and want to come to my house to swap games.

When I first started, I would be the one calling people and asking whether they were interested in doing a trade. In a relatively short period of time, I became the biggest pirate in our local area and had the broadest catalog of software available.

Suddenly, I was in high demand. I had levelled up to become the go-to guy when it came to games and software.

Being at the top of the tree in our area, I was invited to all the major computing events. Everyone in the community wanted to hang out with me. I had gone from relative obscurity to the pinnacle of my chosen hobby, at a very young age. I was only fourteen.

The feeling was addictive. Every time I had something that others didn't, I would get this adrenalin rush which needed to constantly be fueled. It made me work harder to ensure I remained on top.

But local wasn't enough, I needed to level up again. I wanted to become the biggest in the city!

One thing I realized very early on was that there were two types of pirates. There were pirates like me who would want to trade games and not charge anyone for anything, whether it was a service fee, or even a copy fee. Then there were pirates who would make money from distributing software by charging people a fee, which was usually $3 per copy.

If you were paying for ten or twenty pirated games, which was not unusual for the time, you would quickly build up quite an expensive basket of goodies.

Before the Internet and BBSs, people would advertise in local newspapers. The most popular at the time in Melbourne was called 'The Trading Post', a weekly newspaper which came out every Thursday. In it, you could find anything you could think of. From car parts to pets, boats and caravans to furniture. The Trading Post was the go-to paper for teenagers looking to buy their first car. The listings were extensive.

I would often look under the 'Computers and Peripherals' section of the paper to see what computer parts were on offer. People would sell computers, disk drives, printers, monitors, blank disks. Anything you wanted could be easily sourced.

Every week, I noticed advertisements for games. It was always the same phone number, and the list of games was extensive. It was obvious these games were pirated. There was no way someone would be selling games for a fraction of their retail value.

The games listed for sale were also new, which was even more confusing. Why would someone buy an original game and sell it the following week for a few bucks?

I turned to the community to figure out what was happening. As I suspected, these games were pirated copies. Some guy called 'Ginnie'. He was well known in my circle. A 13-year-old kid living in the leafy bayside suburb of Elsternwick in Melbourne's south-east.

Ginnie was making a small fortune selling games to cashed up middle-aged men needing to appease their spoilt kids. He was making so much money from his activities, it was rumored that he paid for a trip to China for him and his family from the proceeds of his lucrative side-hustle!

People I knew had been to his house to purchase games. They described the entire experience as highly organized and business-like. Almost like the 'Soup Nazi' from Seinfeld.

Ginnie had a very professional set-up in his bedroom. He had multiple floppy disk drives all concurrently copying games for his customers. Columns of cases containing thousands of disks were neatly organized and stacked to the height of his room all the way to the ceiling.

There was a back catalog which catered for the older games. But most people were there for the new stuff. The games that were being promoted in the glossy magazines that week. Sure enough, Ginnie had whatever was on show.

He was so popular that you needed to make an appointment just to see him. Thirty minutes at a time – minimum booking duration. Only bulk sessions were permitted, as Ginnie didn't waste his time with ones and twos. Each game would take less than a minute to copy.

During busy sessions, Ginnie had two or three people in his bedroom at once. Outside his room, the next appointments would wait patiently.

The waiting room was like a doctor's office; magazines were available on the coffee table, and his mum would offer hot beverages and cookies while they waited for her son to call them in. It was clinical in its operation.

It was a cash cow. Insatiable demand, unlimited supply, and almost zero overheads.

How could an operation so brazen not come up on someone's radar? Surely this was a slam dunk for any officer with half a brain and an ounce of initiative. What were the games distributors doing allowing this kid to eat their lunch?

I couldn't understand how this was allowed to happen. And it kept on happening every week for years.

Moreover, this oversight by the authorities and content rights owners basically said that it was OK to pirate software. If they weren't interested in shutting down such a brazen and public money-making venture, why the hell would they worry about some guys getting their kicks distributing games for fun?

I took that as a 'Green Light' to keep going. Had the authorities cracked down on people like Ginnie, things would have definitely turned out very differently for me and many in the community.

Personally, I always refused to charge a fee for pirated software because I believe it's unethical. I was a collector as opposed to somebody trying to make fast money out of it. For me, it was a way to learn more about the industry and computers.

It was a passion, not a business.

Being a pirate became an important part of my life. However, being a computer hobbyist was what I really enjoyed. When I received my Apple II computer as a 14-year-old, I was obsessed with learning everything and anything about it. Software and hardware, how it worked, what made it tick, and most importantly how it could be manipulated.

Remember, computers are stupid; a tool we can control to do whatever we like.

I needed to know exactly how it worked from the chipsets to the disk operating system, to the cables, to every single component within the computer; I needed to know every aspect of how it ticked.

Through my network, I would get access to reference manuals. I would read them cover to cover to understand how the hardware worked, how the CPU worked, and how the architecture of the motherboard worked.

In December of 1985 as a 14-year-old, we went away on a five-day summer vacation to my uncle George's house in Philip Island. Everyone had packed their swimming trunks and beach gear for the vacation. I brought along a copy of the Apple II Disk Operating System Reference Guide.

I spent the whole five days nonstop reading this book cover to cover to understand exactly how a floppy disk worked on an Apple II computer. How information was stored, what a 'sector' is, what a 'track' is, what a 'catalog' is. I wanted to understand every aspect of storage on a floppy disk, right down to the hexadecimal level, to the ASCII level, how the drive heads worked, and how things were written onto and erased from a disk.

This fascinated me. I was just obsessed with building my computer knowledge. When it came to computers, I had a deep desire to understand everything about them, not just the software or the application layer, but everything that sits underneath.

This was not only important during my teenage years, but also today in my working life.

Learning about the disk operating system (or 'DOS' as it is often referred to) and how it worked was fundamental learning for me. And being a programmer from the age of twelve, I needed to understand how memory worked, how these memory chips can store applications, and how these applications can fire up and shut down depending on what commands you gave them.

Armed with this new-found knowledge, the very first DOS-based application I wrote was a virus. I wanted to see if I could pull together all the skills I had learned over my brief computing life to create something new.

I spent the weekend writing a virus for the Apple II computer. The virus was specifically designed to corrupt the catalog track of a disk and display a message to the user. Basically, if anybody needed to look for their files, they would not be able to find them because the catalog track would be replaced with my own personal message which would flash across the screen.

I wrote the code in hexadecimal, using ASCII codes, combined with the DOS reference manual I had learned over the summer. I also wrote an antidote, figuring I needed to reverse the damage my virus would inflict.

In the end, I managed to write a very rudimentary, basic virus that would load from a disk into memory and then would copy itself randomly onto catalog tracks of other disks which were inserted in the disk drive.

I decided to test it out in my school computer lab. I placed a few disks around the lab (at this stage I had plenty of spare disks available) and wanted to see how it went.

Within a few days, many of the disks circulating in the lab were corrupted. I was so proud of myself at the time. I had built something using the knowledge I had developed over the previous years and executed the plan flawlessly.

Or so I thought.

Whilst I had the antidote, I hadn't calculated the extent or speed to which the virus would spread.

Once the word got out that there was a virus floating around the lab, the last thing I wanted to do was ride in on my white horse and save the day. I might as well walk into a firing squad blindfolded and ready to take my medicine.

So, I shelved the antidote and kept quiet.

Whispers echoed through the halls of the school as to who was responsible, but no one could prove anything.

In hindsight, it was a dumb thing to do. People frown upon that sort of stuff these days, and I would definitely be annoyed if I was a victim of it.

But this was an exciting period in my life. As a young kid learning the ropes, tinkering, and understanding how computers worked and how everything came together was very addictive. Being able to do the work, research, execute, and then get an outcome that you expected to get, was extremely satisfying.

I went on to write other basic applications, however I was always more of a generalist.

Yes, I could cut code, and yes, I could write some machine code in hexadecimal. I could write source code which would be converted to machine code using compilers. But I didn't consider myself an expert coder. I didn't go too deep into it, although some people might think that I did.

I personally didn't consider myself an expert because I didn't reach an advanced level, which involves manipulating graphics, sound, animation, and other things that good coders can do.

My true passion lies in connectivity, networks, and infrastructure. This became clear to me once I discovered a new and exciting connected world through Bulletin Board Systems.

On Your Bike

There were various ways to access software. The easiest way was to get on your pushbike and ride around the neighborhood.

You would visit friends and computer hobbyists in the area who you knew had a computer and software at home. Some of them didn't have a home computer, but they had software. You would go to a person's house, exchange disks, ride home, copy the disks, and then ride back to exchange the copies.

A very crude method, but that's how it was done.

I had a good friend named Frank who lived in Thomastown, a few streets away from me. Frank also had an Apple II clone computer at home and shared my obsession.

We both met at the Thomastown High School computer lab and established our friendship. He and I would often catch up at each other's houses and swap games.

To be fair, Frank had a lot more knowledge than I did. He was a much better coder technically than I was. He enjoyed going deeper into hexadecimal code and machine code to manipulate programs.

"Well, he is one year older than me, so he should know a little more!", I thought.

However, Frank was a terrible networker. He was an introvert and didn't like the attention or spotlight. Whenever he was in a group environment, Frank would freeze. He was much more comfortable talking to computers than real people.

A typical Saturday afternoon would involve calling Frank to make sure he was home and had time. He always did. I don't think Frank had much else going on in his life.

So, I'd grab my box of floppy disks with the new games I had accumulated that week, jump on my bike, and ride to his place.

It was always a joy to ride along the streets, disks clinking in a plastic supermarket bag hanging from my BMX handlebars.

Whenever I rode to someone's house to swap games, I would get excited. There was this anticipation that I was about to get something new.

Breathing in the fresh air as I rode north on Edgars Road, a light refreshing breeze caressing my face and limbs as I slowly turned west up Barry Road. A few sharp turns later, and I'd arrived.

Frank was a very cautious person. He would greet me at the door, nervously look around to make sure the coast was clear, take my disks, then disappear into his house. I would need to wait on the front verandah while he copied the games. I would hear his mum yelling at him in a Slavic dialect, guessing he was getting interrogated.

"Who's this kid standing on our front porch?".

Sometimes, she would emerge from the front door and offer me a cool refreshing beverage. Summer in Melbourne can be scorching hot, so her kind offers were quickly accepted with much appreciation.

Once I was done at Frank's house, I'd jump on my bike and ride to the next house, this time off Victoria Drive Thomastown. Same deal as before. I'd hand over the loot, and ten minutes later would receive fresh content. Next stop, Station Street Lalor.

This went on for hours, and I loved it! By the time I was done, I had amassed all the new games of the week through my local network.

Everyone I knew was obsessed with games, especially fresh new releases. So, as long as I had access to them, I was cool.

When I attended swap meets, which were held in various parts of Melbourne, I didn't drive. I'd get my dad to drive me there and drop me off. I'd set a time when he would pick me up. While attending these meets, I'd network with people, exchange contact details, swap games onsite, and organize catchups with people at their houses.

We used to arrange private gatherings at someone's house as a "meet". Typically, between three to five people would attend the invitation-only event. The host would have their home computer set up and ready to copy games. Guests would arrive at a pre-agreed time and start swapping games with each other. The host would use his computer to copy the software. Meets would usually go on for hours at a time.

At a meet, most of the times, you flicked through boxes of floppy disks looking for something you didn't have in your collection.

You would get enthralled by discussions around technology and what was happening in the computer hobbyist scene. Alternatively, you'd read magazines and marvel at the latest technology being promoted in the glossy two-page spreads and inserts.

In 1987, the biggest innovation in the home computer market was the launch by Creative Labs of a dedicated audio card, which was later named 'Sound Blaster'. Founded by Sim Wong Hoo, the small Singapore-based electronics company would revolutionize the entire home computer market.

Prior to Sound Blaster, computer sound was limited to a series of synchronized beeps, squeaks, and clicks. Suddenly, the world of computer-generated high-quality audio was introduced to the masses. Creative Labs became so big that by 1995 around 70% of all audio sound cards sold globally were Sound Blasters.

Conversations about technology and trends would fill the void between copying sessions. Often, the host would bring out refreshments and snacks between breaks.

'Radioactive' was a regular at the meets, and he was always open to hosting catchups at his place. He lived in Pascoe Vale South in Melbourne's northern suburbs with his folks and his sister. He was obsessed with Arnold Schwarzenegger and had posters of 'The Terminator', 'Commando', and 'Predator' movies on his wall.

His room was large enough to host five people, and he would always provide snacks and drinks for his guests.

He was also a bit of a gym junkie; he would have dumbbells, a bench press, and a sweaty gym bag next to his bed. A solid build, he was over six feet tall and had short straight dark brown hair, neatly combed to one side, and held in place with hair gel. His square jaw and buffed biceps ensured he was a hit with the ladies.

I always enjoyed catching up with Radioactive. He was from a Greek immigrant family and shared similar values as I did.

He was in his early 20s and had his own car. He and I got along well, regardless of the six-year age gap between us. I was used to being the youngest person at catchups; most were well into their 20s or 30s. But Radioactive always made me feel welcome and, more importantly, treated me as an equal. Our hobby was computers, and we were all excited to share ideas and knowledge.

He was always the perfect host. Generous, funny, and very smart. I learned a lot from him.

Luck's a Fortune

Technology constantly evolves and improves. In 1987, my Apple II clone computer was pretty much obsolete. IBM clone PCs had started penetrating homes, and the Commodore 64 had been in homes for 3 years and was extremely popular amongst the "K-Mart" crowd.

Then, a new computer was released, which was probably one of the most revolutionary of the 1980's: the Commodore Amiga.

The Commodore Amiga 1000 was the first product to be released, and it was a game-changer. The product included amazing graphics, sound, processor speed, memory, and a built-in 3-and-a-half-inch floppy disk drive. The system included a mouse and had a graphical user interface with on-screen icons and folders.

It was truly a state-of-the-art computer which was far superior to anything else on the market and became a highly desirable product. However, the high price tag made it prohibitively expensive for the masses.

As teenagers with plenty of time on our hands, after school, we would catch a train into Melbourne's central business district (or 'CBD'). We would frequent the large department stores every week which stocked the Amiga. We would take turns playing games on the computer, discuss technology with the salespeople, meet up with fellow hobbyists from other parts of Melbourne, and later grab a meal at the local McDonalds on Swanston Street before heading home.

In late 1987, Commodore released two new products: the Amiga 2000, which was an upgrade to the 1000, and the Amiga 500, which was the entry-level model.

The Amiga 500 came with 512kb built-in RAM standard, with an optional 512kb expansion card. This took the capacity to an impressive 1-megabyte RAM, which was an amazing amount of memory at the time. It also included a built-in 3-and-a-half-inch floppy disk drive. The Amiga had superior graphics capabilities, with a whopping 4,096 color palette, dedicated stereo sound chipset, and a high-powered CPU.

In December 1987, after lobbying for several months, I convinced my parents to commit $1,450 to buy a Commodore Amiga 500 with the optional 512kb expansion memory and a dedicated 14-inch color CRT monitor. I managed to rope in another two friends to buy their own units, which allowed us to get a bulk discount on the overall purchase. We were buying three computers, not one. The price had to be competitive!

The Amiga was a huge quantum leap from the Apple II, which had a monochrome green display, a single processor, poor audio and video capabilities, and only 64kb RAM. It was also far superior to the IBM computer at the time. It had a beautiful graphical user interface, very fast processor, and three dedicated chipsets for audio, video, and memory management.

The Amiga games were simply the best you could play on a personal computer, which made the Amiga 500 a highly desirable product that everybody wanted.

Just having one of these amazing computers, you elevated yourself to the most popular kid on the block.

All my connections I made through the years had upgraded to the Commodore Amiga. So, we all started collecting games and software programs that were available for the Amiga.

And again, through the same process of attending swap meets and networking with fellow hobbyists, I built out a strong and vast network of people to swap games with.

I quickly managed to amass a catalog of several hundred games across over 100 floppy disks. The Amiga had a built-in 3-and-a-half-inch floppy disk which had around 880kb storage capacity. This was a huge improvement from the now obsolete 5-and-a-quarter-inch floppy which had around 165kb of storage.

A few weeks after we bought the Commodore Amiga, I met with some friends at a swap meet who explained that they just bought a dial-up modem. This allowed them to communicate with other computers using a telephone line.

Specifically, they were connecting to these things called 'Bulletin Board Systems' (or 'BBS'). Once logged on to these systems, they were able to chat in real-time with people who were also connected.

This was something that had never been available to the general public. At the time, personal computers were standalone units that you used at home or school, but they never spoke to each other, and you certainly couldn't speak to anyone through a computer.

This innovation really fascinated me; the idea that a computer can use a telephone line to talk to another computer was mind-blowing for a 16-year-old teenager.

I would fondly remember the movie 'War Games', a classic early 1980's flick about some geeky kid accessing a supercomputer in NORAD and kicking off a computer game which almost led to global thermonuclear war. By day, David Lightman was an introverted awkward teenager, walking through the hallways of life unnoticed. By night, he had an obsession for computers and games, hacking into games companies to download the latest releases.

They say art imitates life. In my case, I wanted life to imitate art. I wanted to be David.

"How cool was that movie!", I thought.

I realized I needed to level up. This was something I desperately needed to have.

Little did I know at the time that this *exact* moment would mark the dawn of the connected world. Deep down, something inside me knew it was special.

I just had to figure out how to pay for it.

Hitting up my parents for more cash was going to be very tough. They had just shelled out $1,450 for my latest computer gear a few weeks earlier, the last thing I could do was ask them for more cash to buy another piece of kit. As supportive as my parents were for my passion, I suspect even they would have drawn the line at that point. There had to be another way.

I was at school one day when, for some reason, I had a premonition that I should play the lotto that night. I had no idea why, I just felt like I needed to play.

Perhaps images of the modem I desired were dancing in my mind causing me to consider any possible option to get the money to buy one. I didn't want to miss out on the opportunity to join my friends on the BBSs. My gut was telling me I was missing out on something revolutionary.

I needed to have it. I needed to level up.

I caught a train home, and my dad picked me up from the station. I asked dad if he could take me to the local store across the road from the train station so I could buy a lotto ticket.

As a 16-year-old high school student, I had no money. My dad would need to spot me the ticket cost. He agreed, so I filled out the ticket by hand and fed it into the machine.

That evening, I watched the live draw, my one lucky ticket in hand. My golden ticket. I grasped it like Charlie did in a scene out of the classic 1970's movie 'Willy Wonka and the Chocolate Factory'. Hope filled my heart.

The first number fell, and I had it. The second number fell, I had it. Before I knew it, I had all the first four numbers, and I was getting excited.

The fifth number fell, and I had it. OK, I thought, "This is unbelievable!"

The sixth number dropped, which I didn't have.

That was it. No more hits. Game over.

So, I got five straight numbers in the lotto that evening. My dad looked through the ticket and realized that I had played the sixth number in the first 10 of the 12 boxes, but not in the last 2 boxes, which is where I hit five numbers straight.

Hitting five straight numbers in the midweek lotto draw paid roughly $1,500. It was the closest I'd ever come to winning the lotto jackpot, not that I regularly played the game before or since. In my mind, lotto was a tax for the stupid. The odds are stacked so far against you, it wasn't worth the price of admission.

But at that moment, $1,500 was a great return on a $10 investment.

I gave my dad a thousand dollars, thanked him for bankrolling the ticket, and I kept $500. It was a good deal for me, and a good deal for him. He had recouped most of the money he paid for the Amiga a few weeks earlier.

I immediately went out and purchased a Bit Blitzer modem with my $500 windfall. The modem price happened to be exactly $500. That was all I needed, so I was happy.

The modem had a matt black grainy metal case with red and green flashing LED lights on the front. Each light represented various modem functions: Send Data, Receive Data, Connection Status, etc. It was a beautiful looking modem and was one of the best rated on the market at the time.

It had dual transmission speeds: 300 and 1,200 bits per second send and receive speeds, built-in error correction, and auto-synching. Basically, it was so sophisticated that it could connect with any other modem on the market and stay connected even if there was severe line noise. It was rock solid.

So, I bought my first modem in early 1988. I was in Year 12; I was 16 years old. I plugged the modem into my Commodore Amiga, and suddenly, I was connected.

I had leveled up.

Now, being connected back then was not the same as today. In 1988 there were only two types of systems you could legally connect to: University computers or privately operated Bulletin Board Systems (or 'BBS').

Bulletin Boards were the precursor to the Internet that we all know today. They were effectively hobbyists that would set up a home computer running off-the-shelf BBS software, connect the modem to it, and have a telephone line coming into that modem.

This would allow other computers to connect to the BBS and access the specialized software that ran on that computer, facilitating user chat services and, importantly, a file-sharing system.

HACKERS, CRACKERS, PIRATES AND PHREAKS

A New Online World

I'd discovered this exciting new world called Bulletin Boards. However, I quickly realized that this world was not really "open to the public".

Most of the BBSs were restricted access, reserved for only the chosen few. Accessing them was a privilege, not a right. Just because you had a modem didn't give you automatic access to these privately controlled systems. You needed to know the System Operator (or 'SysOp' as they were referred to), who was the owner of the BBS. You needed the phone number to dial into the BBS, and you needed a user account.

This was familiar territory to me. It was the computer lab all over again. I understood the dynamics at play, but this time I was much better prepared for the challenge.

Only the SysOp could grant you a login. Although most BBSs allowed you to create a new account on their landing page, this would inevitably need to be approved by the SysOp. If you were new to the community, or unknown, you wouldn't get in. The only way to fast-track this process was to either call the SysOp (because you knew him) or get a friend to vouch for you. It wasn't done remotely unless you were known in the community.

Smaller BBSs were a lot easier. They would allow anyone to get an account and access their service. Most of these were run by single mid-20s males from their parents' house as a way to tinker with technology while their mum cooked dinner. The popular ones, however, were tough to get on.

Therefore, you really needed to know the person who owned the board, or you needed to be introduced and vouched for by someone within the community. I was fortunate enough to already be entrenched in the Melbourne computer hobbyist community, and therefore managed to get onto two of the most popular BBSs in Melbourne.

One was called Pacific Island, and the other was called Zen.

The main difference between Pacific Island and Zen was that Pacific Island had a single dial-in modem where you could leave messages on the bulletin board and share files.

Zen, on the other hand, had four phone lines connected to it via a rotary system, which meant that up to four people could be on the system at any one time. Zen allowed you to chat in real-time to other users who were also logged into the BBS at that time, leave messages, share files, and access both public and private chat rooms.

That was a huge advantage. Remember, this is 1988. The Internet as we know it did not exist. Connectivity was in its infancy. This ecosystem was being built at a rapid pace by computer hobbyists; trailblazers who contributed to what eventually became the Internet.

The process of logging on to Zen was relatively straight-forward. Using specialized modem software, you would dial the telephone number to the BBS. When the computer answered, a unique-sounding series of unsynchronized squeaks and crackles began trumpeting between the two computers. The symphony to align both modems had begun.

After a few seconds, both systems would screech the same unified noise; they had finally agreed to talk the same language. After a few more seconds, the login screen would reveal itself.

Slowly but surely, every line on the screen would sequentially be populated with the characters of the login page. A series of white text on a navy-blue background that, when combined in a certain order, revealed an image of a large heading.

"Welcome to Zen". In big text-based lines.

All the usual "prohibited access" warnings would grace the footer. Someone had spent time designing the characters to perfectly line up with the standard 80 x 25 text-based screen matrix.

Finally, the dance was over. Time to login.

Bulletin Boards started off innocently enough. A place where people could meet and share information about technology, life, love, and anything else that was of interest. It was a place where the misfits would congregate. People shunned by the real world who found a place of solace which allowed them to express themselves openly without fear or favor.

In the real world, we were labelled nerds. A name popularized by Hollywood movies of the day. Geeks. Freaks. The guys who got beat up in gym class. The stereotypical weakling who had brains not brawn.

The reality was very different. Many of the people I met in the community were the complete opposite to what you'd expect. Some were addicted gym junkies. Others were sports athletes. Others had girls chasing them for dates. The one thing which bound us all together was our curiosity for technology. A thirst for knowledge that wasn't being fulfilled by the education system.

I remember a time in high school where I witnessed these two worlds collide. I was in Year 9 at St Johns, and we had physical education that afternoon. I had brought in my Vic 20 tape drive for a friend of mine who had a Commodore 64 and needed to store some files.

During my high school years, I was quite active. I simultaneously played basketball and outdoor soccer, and trained karate. As a result, I was quite fit and reasonably well defined. I enjoyed team sports. It was a good way to meet people and socialize.

Dressed in my basketball singlet and tight shorts, biceps flexing and leg muscles shimmering, I had two girls approach me from the class. They were obviously attracted to the attire I was wearing. While chatting to them, I leaned over to grab something out of my bag when one of them noticed the tape drive. She immediately recoiled and started laughing, pointing to the foreign object. Then she cried out with a flirtish chuckle "Oh my God! You're a NERD!!".

I had quickly transitioned from a cool athletic kid to a geek in under 3 seconds. My hobby had been discovered and screeched across the classroom. Students hovered around me to see what all the fuss was about. Mostly girls.

"They like to hunt in packs!", I thought, as I quickly shuffled the contents of my bag to hide the tape drive.

Whilst this didn't necessarily affect my social status amongst the other students, from that point on I was known as the computer nerd by the entire student body. The brainiac when it came to technology. A Hacker. Someone who would rule the world someday.

The branding was so widespread that I often had teachers approach me to ask for help with their own computer projects. Usually when a teacher calls you out during lunch time, it wasn't good. Detention would almost always follow.

In my case, they would pull me aside and quietly hand over a floppy disk of something they were working on and needed help.

I became the go-to guy when it came to technology for both students and teachers at the school.

A NEW ONLINE WORLD

As a member of the BBS community, I would often get invited to "catchups"; a social gathering for users to meet up in person. It was an opportunity to get to know the people behind the online personas. A way to exchange stories and ideas without the restriction of a keyboard.

Catchups were usually held in the central business district of Melbourne. They were usually held once a month on weekends, as most of the attendees were high school students like me.

I was 16 years old when I went to my first BBS catchup. I got dressed up, 'Euro' style, not knowing what to expect. I wore dark-blue Levis 501 jeans, a black short-sleeve Gucci sweater with a massive white logo across the chest, and black suede leather lace-up boots. The outfit was topped off with a full-length black trench coat; collar flicked up to make me look more mysterious. Pino cologne was splashed across my face and clothes, giving out a unique and overpowering Italian playboy scent.

I caught the train to the city and went to the designated meeting point: City Square, on the corner of Swanston Street and Collins Street.

The City Square was *the* meeting place for 'cool' 1980's Melbourne teenagers. The fully paved open space measured roughly the size of a football field and ran the entire city block from Collins Street to Flinders Lane. To the left was a ten-foot-high concrete and corrugated tiled wall which was around 30 feet long. Water would trickle down the front of the wall into a fountain. In front, concrete balustrades would be used as either seats or ramps for the skateboarders.

Central Station Music was located in the center of the City Square. Every Melbourne teenager would flock there to listen to the latest vinyl albums from the biggest pop icons of the day. It was also a great place to meet girls.

We would always meet on the north-west corner opposite the Melbourne Town Hall, a grand two-story bluestone period building built during the 1850s Victorian gold rush boom.

From there, we would usually either hang out at a local game arcade like Ten-Four which was located on the corner of Swanston Street and Little Bourke Street. We'd drop coins in machines and play the latest games of the day, such as Dragons Lair, Outrun, Afterburner, Double Dragon, and Gauntlet.

Other times, we would go and watch a movie, usually Action, Sci-fi, or Horror. I avoided Horror movies as they really didn't appeal to me.

Final stop: McDonald's on Swanston Street near Flinders Street Station. This place was a popular hangout for teenagers. It also had another game arcade next to it called Flashback which would soak up our last remaining coins before we headed home by train.

This group of BBS enthusiasts were effectively the first people to go 'online' in Melbourne. Not surprisingly, most of them were of Australian backgrounds.

Now, for those of you who didn't grow up as a teenager in Australia in the 1980's, there was a dividing line between the immigrants and the local Australians.

Specifically, there was a separation between the 'Aussies' and the Europeans. The Aussies used to call us 'Wogs', which meant a virus or disease. We would in turn call them "Skips", which was a popular kangaroo character on Australian television at the time.

The two races never used to mix. All my close friends were born in Australia but were of European migrant parents.

And so, even being in the same group as these guys was very strange for the time.

To my surprise, this group of people accepted me in their circle with open arms. All my preconceived concerns went out the window. At their core, they cared about the same things I cared about: Technology, computers, connectivity, and sharing information.

There were many characters who attended these meets. One in particular went on to become notorious internationally.

His name: Julian Assange.

Julian was part of the Melbourne crew. We used to both access Zen and Pacific Island. He was a regular at the Melbourne CBD catchups, and we met several times. We're both roughly the same age and shared the same passion for technology.

For us, it was all about witnessing the evolution of computers from a stand-alone tool to a connected platform. All these changes were happening in front of us, and we had a front row seat to watch the innovation unfold before our very eyes.

The world as we knew it was changing.

At the time, Julian was very much anti-establishment. In political terms, I would say he was a bit of a leftie, very righteous, and always looking to stand up for the little guy. He truly believed that the entire system was corrupt and would come crashing down one day, and technology would enable people to bring the system down. I remember some of these discussions taking place over a Big Mac at McDonalds Swanston Street in Melbourne in 1988.

I suspect Julian probably wouldn't remember me. To be fair, I barely remember him. However, as soon as I heard the name Julian Assange and Wikileaks, I knew instantly that it was the Julian I knew from back in the day.

At the time, Julian was dating a girl whose online pseudonym was Electra. From memory, she was a sister to one of the SysOps and was one of the very few girls who used to regularly come to the CBD catchups.

Most of the group were like Julian. They would listen to Midnight Oil songs about nuclear weapons and US military bases, and about how mining was destroying the native land we were on. Powerful vocals for a teenager searching for a meaningful cause.

They would also imitate the lead singer by wearing flannel shirts, white T-shirts, blue jeans and black or tan boots. This seemed to be the uniform of the Aussie underground community of the day.

Girls like Electra were more 'Goth', with black fishnet stockings and ripped shirts. Black was the color of choice with a sprinkling of white for contrast, and plenty of black mascara and eye liner.

I was clearly over-dressed when I first attended.

"Next time, I will tone down my dress code…", I thought.

The group were tight, not allowing anyone else in without a nod from their own. They partied together, travelled together, and shared things together. They had certain rituals which they followed.

For example, they would regularly go to the Westgarth Theatre, located on High Street Northcote. Opened in 1921, the theatre remains the oldest purpose-built cinema still operating in Melbourne, receiving a 'Local Significance' classification by the National Trust of Australia.

And you could tell. Stepping into the foyer was like walking through a time warp, emerging in the roaring 1920's.

The fixtures matched the overall allure, with a splattering of Art Deco style plastered across the ornate ceilings and timber wall skirtings. The convenience stand was made of solid mahogany wood stained in a dark brown finish, history etched in the timber paneling from all the bumps and bruises of a century of loyal service. Treats and snacks lined the front counter, brightly lit mirrored shelves on the back wall displayed bags of potato crisps and candy, and the sweet smell of freshly cooked popcorn slowly billowed out from the red and yellow popcorn maker, enticing you to buy a bag for the show.

You could feel the pulse of yesteryear as you walked up the wide center staircase leading to the cinema which was small and cozy. Perfect for nostalgic movies which had entertained the masses all those years earlier.

The air in the cinema was thick and still, with a dusty fragrance. The theatre seats were vintage thick cowhide burgundy red leather, each with a single heavy-duty spring in the center of the cushion. Whilst they had certainly seen better days, they were functional and added to the overall atmosphere of the historic complex.

The group would congregate at the 'Valhalla', as the Westgarth Theatre was called, for the regular mid-week screening of the 'Blues Brothers'. They would all get dressed up as the movie characters; most wore black suits, black fedora hats, black ties, with a white shirt and dark sunglasses. They would recite every word of the movie and laugh loudly at the funny scenes. When the cars hit the newspaper stand in the mall chase scene, newspapers would be thrown across the theatre to rapturous applause. When the car fell into the swamp, water pistols would squirt out to the patrons.

It was a 4-D experience like no other.

Other times they would attend the screening of 'The Rocky Horror Picture Show', another classic movie of the day which drew a big crowd. Same deal as with the 'Blues Brothers' film. People would come dressed as characters from the movie and recite the lines and the scenes in real life.

Electra and other girls in the group (there weren't many) would often invite me to attend these events. She would explain how much fun it was and how everyone got into the spirit of the moment.

The more she explained it, the more I realized it wasn't for me.

I was not really interested in this side of the group. After all, we 'Europeans' had too much style to do this sort of stuff, right? My nights were better spent attending street racing and burnout meets around Lygon Street Carlton, the center of the city's Italian community and a great place to hang out and meet girls.

Many of the people who attended these Melbourne BBS meets went on to become infamous.

Importantly, this was where I first met Craig Bowen.

Craig was the owner and SysOp of both Zen and Pacific Island. He was Numero Uno. The main man. The guy everyone wanted to meet.

His online presence was legendary. Every time you logged onto one of the BBSs he operated, you would send out a call for the SysOp to come online. Sure enough, day or night, Craig would promptly respond and jump on a live chat. He was elusive. He would only jump on if he knew you personally or if you had been vouched for.

In real life, Craig was not what I expected. A slim late-20s male who was extremely quiet and reserved. Flannel shirt, check. White T-shirt, check. Blue jeans, check. Tan boots, check. He had thinning chestnut hair, which was cut short and neatly combed to one side, like a kid in one of those nerdy 1970's photographs. He had an oval head which would make a great stencil for a 'Humpty Dumpty' caricature.

Craig wasn't the type to drum up a conversation. His modus operandi was to sit back and absorb the discussions from the teenagers around him. Like a parent making sure their children weren't doing anything stupid.

He would often say that he was there to make sure we behaved. I suspect he enjoyed the social aspect of the meets. I didn't get the feeling he had much else going on in his life at the time.

From memory, he was a Telecom Australia technician of some description, which explained how he was able to get five phone lines installed in his bedroom to run the two most popular BBSs in Australia.

I always found it strange that a guy his age and personality enjoyed hanging around with a bunch of adolescent misfits reaping havoc across the Melbourne CBD.

And he had good reason to be watchful. This group of misfits would do a lot of dumb things.

Craig would constantly be corralling the group to make sure none of the flock drifted away. The local McDonalds would get a special baptism of fire when we swarmed in. Pickles from the burgers would get flicked on the walls, sodas would be splashed across tables, and meal trays would be stacked up to form towers which would then be demolished with a flying karate sidekick.

The poor McDonalds workers would then reluctantly go around scraping the pickles off the walls and wiping down the tables while the misfits looked on in rapturous laughter.

"Wogs would never act like this. Our mums would kill us!" I thought.

But when it came to technology, particularly hacking, Craig was one of the best. He had access to all the reference manuals from his work at Telecom Australia and was well versed in the innovations in the industry at the time.

Craig was plugged in, connected to the global computer underground like no-one I'd met before.

Interestingly, 'Craig Bowen' was not his real name. He was the only guy I knew who had a pseudonym for a pseudonym for an online persona.

We all knew him as Craig, and that's just the way he liked it.

HACKERS, CRACKERS, PIRATES AND PHREAKS

Going Global

The first thing you learned when you were involved in the community was to never use your real name in anything you did.

You needed to use a pseudonym which was effectively your call-sign or what people referred to you as. Basically, you had to come up with something unique, that was not used by anybody else in the community, which reflected who you are and what you did.

At the time, this was a new concept for me.

During my early pushbike "door-to-door" pirating days, I used to use the name "Cobra Strike". Mainly for two reasons: I liked the name 'Cobra' (it sounded cool), and there was a movie out at the time with Sylvester Stallone called 'Cobra'.

However, I realized I needed to separate my pirate name from my online name, so I had to come up with something different. I came up with the pseudonym "The Rebel".

It was not like I was breaking any rules or doing anything illegal, but I felt Rebel was something I could align with.

For me, being online meant you were being a little rebellious. You were doing something that others weren't doing, didn't understand how to do it, or weren't connected with the right people to do it.

All my life, I have always been the kind of person who likes to do things differently, not wanting to necessarily follow what other people are doing. But at the same time, from the grounding that my parents gave me, I had a clear sense of right and wrong.

Now, I know people will say that pirating is illegal. But again, for us, pirating was not to profit, but to share and learn and grow. We saw ourselves as collectors, so we certainly didn't believe we were breaking any laws per se.

And always in the back of your mind, you kept thinking "I'm under the age of 18, so if I do get caught, what are they going to do to me? They might confiscate my disks and slap me on the wrist".

Every school in Australia had pirated software circulating in their labs; it was normal behavior at the time. So, we really didn't worry about it.

Everyone online had their own unique names, and some were quite funny. There was Bit Mapper, Interceptor, Killer Tomato, Ivan Trotsky, Blue Thunder, Electra; all kinds of names were being used.

Before bulletin boards, most people in the community were pirates or technology enthusiasts. But when bulletin boards came into existence in the mid-1980's, many pirates began transitioning into hackers. The community originally focused on swapping software and collecting games, which gave you a bit of status and notoriety amongst your peers.

Suddenly, a new connected world began revealing all its wonders, seducing the community.

Many of my closest contacts I met on the boards began the metamorphosis to hackers. And being part of this closed tight-knit community, of course, you tend to follow what everyone else was doing.

I remember chatting to a few hackers on Zen to pick their brain on how they got started and what they were up to.

Like anything in life, you need to learn to crawl before you can run. Hacking was no different.

First, you gained access to an admin account of an "easy" target, usually a local college or university computer. These credentials were passed on by experienced hackers to those interested in learning the ropes, kind of like a teacher giving you access to a sand box so you can have a play. You would poke around and see what files were available, check the system logs and folders, set up new user accounts (preferably with admin rights), review the email system, and start to understand the basics of moving undetected across a restricted system.

Once you mastered the first system, you would get access to a second, and then a third. Building your knowledge base over time, you started to understand how hacking works, how systems work, and how they can be manipulated. Importantly, you started to appreciate the value of information.

The community was always a great resource for learning and would often help interested members gain valuable knowledge and experience. Over time, you would gain access to reference manuals and syntax glossaries which would come in handy during your progression through the hacking ranks.

At this point, the training wheels were off, and you were on your way to doing your own exploits. Before you knew it, you were targeting your own system and using the tools at your disposal to break in.

The first time is always special, regardless of which system you break into. It's like your first love; you never forget it and always remember it with fond memories.

Personally, I found hacking very boring. It simply wasn't interesting. I dabbled a little in it and played around with various university and college systems and networks. But for others in my network, they got an absolute kick out of it. They truly loved it.

None of these activities were discussed on the bulletin boards, at least not on the public sections.

As you meet more people through the online community, you start to build out your reputation. I was especially active on Zen at the time. I used to communicate with a lot of people and would be a regular at all the various Melbourne CBD catchups. And the more I networked, the more I learned. I managed to find people who were involved in the 'inner sanctum'.

Now, the inner sanctum of a BBS was effectively a closed ultra private section of the bulletin board that only the SysOp would grant you access to. This was the absolute crème de la crème, top of the tree, very restricted domain.

Because of my solid reputation, network, and well-established pirating skills and inventory, I was invited to join the private sections of both Zen and Pacific Island.

That is when everything changed.

These sections of the BBS were the exclusive domain of the top-tier pirates and hackers from around the world. Information would be freely shared in this secret area of the BBS.

The things that were available to members were out of this world.

Looking back now, I still shake my head in disbelief as to how the hell I got into these private sections. Into this elite world.

At the time, I happened to be one of the top pirates in Melbourne and had access to a lot of games. The pirate section of these BBSs had more games than I could ever dream of having. It was a much wider catalog of games than what I had in my possession. And it was all stored in these private areas of the BBS that only a handful of people had access to.

I was like a kid in a candy store!

Prior to going online, pirating involved physically meeting with someone and swapping games. Usually at someone's house or at a swap meet. That was the standard way to distribute software.

However, it was cumbersome and lethargic, like a bloated walrus trying to weave its way across land in search of the solace of water.

What I never considered was how the software ended up at the swap meets in the first place.

Remember, there was no online distribution method. Things had to turn up the old-fashioned way: through the post.

Imagine having to wait weeks for a parcel to arrive in the mail which contained new pirated games. Add to that the complexity of finding the international sources who would ship the parcels to you, all for a fee of course. It was a very difficult exercise.

Sometimes you would have to wait for the games to be released in Australia, which was often weeks or even months later than Europe or the US. You then had to hope that someone in the community had the money to purchase the original, and then had the skills and discipline to spend a weekend cracking the copyright protection. Pirated games would therefore take months to land in your hands.

With the dawn of the BBS, suddenly pirated games began arriving almost as soon as they were released. Cracking teams were sprouting up all over the world, obsessed with being the first to crack a new game and distribute it to an insatiable global audience; Their trademark handles and pseudonyms branded all over the introduction screens.

And the best cracking teams resided in Europe.

The number one cracking team in the world at the time was made up of Germans, Austrians, Belgians, and Dutch. These guys would arrange weekly catchups at someone's house and spend an entire weekend cracking newly released games.

The invitation would go out containing the location and time, along with a list of games which were on the menu. Team members would then jump in their cars and drive to the designated city where the cracking would take place.

I often wondered what it would be like to be one of these elite crackers driving along the German autobahn heading to a meet with their fellow teammates. I imagined the anticipation in their hearts and the adrenaline running through their veins. Not to mention the empty potato crisp packs, candy wrappers, and soda bottles thrown around their cars.

They were meeting up with other top-tier crackers and doing important work. They would relish the challenge of breaking the unbreakable. They would turn their minds to disabling the sophisticated copyright protection methods devised by some of the smartest computer professionals in the world.

Challenge accepted!

In all my time working with these teams, I never heard of any game which could not be cracked. What was surprising was the speed in which these copyright protections were being disabled. It's as though the professionals were getting lazy and deploying the same techniques across all titles, making them easy targets for these world-class cracking teams.

I was intrigued to know who was funding the purchase of these games in the first place. You needed to have a copy of an original to crack it, and these things were turning up in large numbers every weekend.

The answer, as I was to find out later when chatting to these guys on a German BBS, was that a few of them worked at a local computer store. As games were released to the stores, they would borrow them for the weekend to "have a play". What they were really doing was putting together the next batch of titles for the weekend feast. Monday morning, the games would be returned to the store, ready for resale.

The formation of the 1980's global computer underground had begun. The pieces were starting to come together. And connectivity was driving its emergence.

I immediately started downloading every single piece of software on Zen that I did not have, which was a lot. I would sit there tying up the home phone line non-stop from 9:00pm until 5:00am. Download after download, every single night, 7 days a week.

But of course, the BBS private sections had a quota system. Basically, you were free to download files and information, but you also needed to contribute to the community. If you blew your quota by constantly downloading files and not uploading anything in return, effectively taking more than what you were giving, the SysOp would start to get a little bit annoyed with you.

I had to figure out how all these games were becoming available on the Australian BBSs. Where were they coming from? Who were the people that were bringing these games in?

In essence, I needed to figure out how I could level up to be one of those people.

And so, through networking with people in the private chat sections of the BBSs, I started to meet pirates that didn't do what I did. They didn't get on their bike and ride to someone's house, or attend swap meets, to swap games. They used their modems to dial into international BBSs from around the world and instantly download games.

Now, in 1988 the Internet as we know it did not exist. Any international phone call would cost in excess of $2.00 per minute. How then were these people managing to dial international numbers and stay connected for hours so they could download games. On the other side, you had hackers, who would hack into computer systems located all over the world.

This piqued my curiosity. I had never seen or read about this before. This was stuff that was not readily available to people in the public domain. You had to be at the top echelon to understand this space.

I decided I had to become one of these people. They were at the top of the global tree, not just the local computer community. Only the best people in the world were doing this stuff.

That's where I wanted to be. That was my new goal. So, I started working very closely with some highly regarded hackers and pirates in the global computer underground which helped me figure out how they were doing it.

There were two steps you needed to take to access international bulletin boards and make free phone calls in order to download software and games.

Step number one: you needed to get verified and validated on the private sections of these international boards, which were reserved for the most exclusive hackers and pirates in the world. Because of my extensive network and existing access to private BBSs in Australia, I was being vouched for to these international bulletin board sites in the US and Europe.

Getting access to private BBS systems was slow going, but the login details started coming through. I got accounts on private sections of BBSs in the US, Germany, UK, and Holland to name a few. All my accounts were top-tier, all-access in private sections of the respective BBSs. Importantly, all these systems had the latest software and hacking information.

Ali Baba had opened sesame.

So, the first part was solved: credible access, login name, and password.

The second part was a little bit trickier. How do you make free phone calls to be able to get onto these systems to download all the games and information.

There were three ways to do it.

The first way was to log onto a central dialing service of a parallel statutory authority to Telecom Australia called the Overseas Telecommunications Commission (or 'OTC'). At the time, OTC were responsible for all international calls made to and from Australia, and had a central system based in Canberra. You basically had to log into their system via a local number and then patch through your connection to the international number that you wanted to dial.

This was a very risky play. You were effectively tapping into the actual terminal screen of the operators who were monitoring the phone lines. If by chance the operator happened to be sitting at their desk when you were accessing their terminal to jump to an international number and they spotted you, they could trace the call back and you were busted. As a minimum, they would abuse you and kick you off their system.

Although the OTC approach was by far the easiest, it was probably the riskiest way to get free international calls.

The second way was to call up a US-based carrier using a calling card, via a local number. When you dialed the carrier number, you would be patched straight through to a US-based operator for the cost of a local call.

The US carrier of choice at the time was AT&T. Their direct number was 88-1011. From Australia, you would dial the international direct dial prefix of 0011.

However, many people weren't aware that OTC had a second prefix which was designated for fax machines. In the 1980's, Australia's international direct dial phone lines were notoriously bad. They contained a lot of noise, which was the worst thing for a modem.

Noise meant the lines would intermittently drop, and you would need to redial the number. A major pain in the backside when your half-way through downloading a game and the line drops. You'd have to reconnect and start all over again.

The BBS inner sanctum had the solution, and it was a golden nugget.

By using the prefix 0014, you would get the best possible line which was available. The reason: The 0014 prefix was reserved for fax machines. By dialing 0014-88-1011, you got a perfectly clear line straight to a US operator waiting to patch you through to your destination phone number.

The US-based operator would ask you to provide your unique calling card number. You needed to put on an American accent and sound like an adult. It took a lot of practice to get the accent right.

Australian accents were easy to pick. We have a way of saying things which anyone beyond our shores would instantly recognize. Calling a US-based carrier and speaking in your native tongue was never going to end well.

When Australians put on an American accent, most of us sound like southerners. Like we're from Mississippi or Arkansas. Australians pretending to have a US accent would usually have a southern drawl which sounds funny to the Yanks. Whilst crude, it was effective. It was good enough to get through the gatekeepers.

But the accent was only half the battle. You needed to sound like an adult. No point calling the operator sounding like an adolescent teenager with a squeaky voice. You needed to put on a low-pitched voice to accompany your fake accent. Tricky certainly, but still doable.

A calling card number is effectively like a credit card number. The main difference is that it is connected to a person's phone bill. Once you provide the calling card number, making sure you sounded like an American adult, you then provided the phone number you wanted to connect to.

Once you were connected to that number, you were through. You could then sit on the line for hours for "free" and download as much software as your heart desired.

The third way was far more sophisticated. Only the top hackers and phreakers in the world knew how to do this.

In the 1980's, many large corporations started installing Private Automatic Branch Exchange systems (or 'PABX'). These systems would be installed at all major offices across cities, states, and countries of a large corporation. They were effectively private telephone networks which companies installed so that they could avoid using the carrier lines for internal calls, thereby reducing their telephony bills.

These PABXs would typically be interconnected via dedicated leased lines between offices, which allowed employees to make calls to each other within the organization without being charged a call fee by the carriers.

PABXs had another important function: they allowed employees to dial numbers outside the company telephone list. Some employees were pre-authorized to dial international numbers. Therefore, PABXs all had dedicated outside telephone lines.

Using a modem to dial into a PABX, a hacker or phreaker would gain access to outside lines and be able to make free phone calls to anywhere in the world.

But accessing a PABX wasn't that simple. You needed to understand how the system was interconnected, know the local telephone number to dial which gave you access to the login screen, have the correct administrator login credentials, and know the commands and languages used by the PABX to be able to maneuver around the system.

Each PABX system also had an automatic log which kept track of all inbound and outbound calls being made. These logs were reviewed and cross-referenced by the corporations and reconciled against their phone bill. This effectively allowed suspicious activities to go relatively unnoticed until the next telephone billing cycle.

With your Administrator login credentials, you would manipulate the logs and make any changes you desired, such as changing the inbound and outbound telephone numbers, or deleting the calls altogether, effectively removing anything which linked you back to the PABX.

PABXs were therefore the preferred method of choice for making free phone calls by the more sophisticated members of the global computer underground.

Accessing an international BBS was basically the same as accessing Zen or PI. The main difference, of course, was that the system you were accessing was located on the other side of the world.

GOING GLOBAL

The whole idea of sitting in your bedroom in Melbourne and logging onto a BBS in Munich via a home phone line was mind-blowing for me at the time. Seeing German words appear across the screen as your modem fetched the data was an experience like no other. It felt exotic, like you had instantly been transported to a foreign country.

There were also other differences. Firstly, some BBSs would display their text in the native language of the site (such as German or French), meaning you needed to decipher the words to basic functions like the menu items, commands, comments, and games catalogs.

Communicating could also be somewhat challenging. Some users would use their native language in the live chats. You needed to explain to them that you were from Melbourne Australia, and only spoke English. Most would oblige by chatting to you in English, albeit broken (their vocabulary was somewhat limited).

While chatting, you would pick up new words used across the global computer underground which were restricted to people in the inner sanctum. The most common word which was used by almost everyone in the community was "lamer".

The word "lamer" first appeared on pirated software in the mid-1980's, although it originated in the early 1960's. It basically means someone who doesn't know what they're doing, or is incapable of doing something. A person with average or mediocre capabilities.

It was used as a derogatory word in the global computer underground.

Anyone outside the inner sanctum was therefore referred to as a lamer.

Whenever we chatted in the private sections of the global BBSs, the word lamer would be thrown around for two reasons. One: it showed you were up to date with the latest international lingo. And two: it showed you were part of the inner sanctum. Only people at the top level knew exactly what it meant.

When you used the word lamer in the private sections of Zen and PI, others would instantly know that you had learned the word from the global computer underground. It was a very powerful word; one which elevated your status in the community.

Somehow, over time, the word began to enter the mainstream. People in the public areas of Zen and PI started using it when abusing another member of the BBS.

What was once a secret word had found its way to the masses. Suddenly, the word "lamer" had lost its luster.

Another word which was very popular at the time was "uber", a German word which means "ultra", "over" or "above". The word was being used heavily, as you would expect, on the German BBSs. It slowly made its way to other BBSs around the world, most likely by the inner sanctum who frequented the international sites.

The word "uber" would often be combined with others to emphasize an activity or description. For example, "uber cool", "uber smart", and even "uber lamer".

Today, the word "uber" has become synonymous with the ride sharing behemoth. I often wonder if any of the Uber founders were active on BBSs back in the day. That would explain how they chose such a unique name for their entrepreneurial venture.

When chatting on these international BBSs, you bumped into a lot of people from all over the world. Whilst English was the preferred international language of the BBSs, this would of course vary from country to country.

For example, anyone from a Commonwealth country would type their words in UK-English. They would spell their words differently, such as "colour" instead of "color", or "organisation" instead of "organization".

In addition, you would identify the nationality of the person by the words they used in their chat. For example, Australians would use the word "mate" which means friend. Americans would use local lingo such as "home run" or "QB".

Every user would also have their own chat vocabulary, depth, and style.

Germans, for example, would type in a very clinical manner, straight to the point. Not much fanfare or excitement. Sometimes, they would throw in a German word in their chats which identified who they were and where they were from.

After spending hours downloading games from various international BBSs, you would upload these games onto other global BBSs. You needed to keep your ratios up on all the sites you visited to make sure you were allowed access to download more games when they became available.

The problem, of course, was that you were often using calling cards, which were linked to an individual's telephone account. These individuals would eventually receive an astronomical bill at the end of the billing cycle.

To try and alleviate their pain, and spread the risk, you would use multiple calling cards and systematically rotate through them.

At the time, calling cards were relatively easy to get. Some people would get them by simply flicking through a US phone book, targeting the Beverly Hills area. This was the most well-known and "ritziest" area of America to a young Australian hacker or phreaker.

People would call a number randomly from the phone book pretending to be a carrier call center operator and ask the customer to confirm their calling card details because of suspicious activity. Customers would willingly give their calling card details over the phone. These were then recorded and distributed throughout the private BBSs globally.

Today, the media is filled with Nigerian Scam stories. People calling the vulnerable requesting payments for something or someone. Text messages sent with phishing credentials, designed to access personal and sensitive information contained on your mobile device. Even Indian scams are starting to grow in frequency. Many would have received a call from an Indian call center claiming to be a Microsoft employee, miraculously offering to solve our impending Windows licensing issues.

The other way to get calling cards was through hacking into the carrier systems and extracting active files containing calling cards. In the late 1980's, computer security was not what it is today, so these types of breaches were more common than most would have thought.

The other thing that was readily available on the private sections of the bulletin boards was credit cards.

Credit cards and calling cards were the two most desirable commodities on any private BBS. If you were in the inner sanctum, you had access to as many as you wanted.

Personally, I had absolutely zero interest in credit cards. It simply wasn't in my nature to commit credit card fraud. The whole idea did not sit well with me.

Calling cards, on the other hand, were different. In our teenage minds, calling cards simply meant free phone calls.

With calling cards, typically, if there is a disputed bill on the carrier side, especially if it was an international number, the carrier would waive the charges to the customer.

And besides, I've never met anyone who actually feels sorry for the carriers. Most customers have a love-hate relationship with them. Mainly hate.

Credit cards were, and still are, a different story.

Nonetheless, you would have access to tens of thousands of valid credit cards and calling cards. If you ever needed any, they were readily available in the private sections of the BBSs.

In today's Internet, there is an area called the 'Dark Web' where similar activities take place. In fact, I wouldn't be surprised if the people involved in the underground BBSs of the 1980's went on to build the Dark Web of the 2000's. The concepts and methodologies are strikingly similar.

Back in 1988, most people had no idea what was happening over their phone lines. The things we were doing were absolute bleeding edge, before the Internet as we know it was invented.

I remember sitting in class as a 16-year-old looking around at my fellow students, thinking to myself "you guys have no idea what's happening under your very noses".

I used to visit people's homes and get drawn to the telephone line. It was magnetic. I remember observing the people in their houses go about their daily lives, completely oblivious to what was happening just a few feet away.

Knowing what I knew, and being one of the best in the world at what I did, there was this sense of power and privilege.

I had no interest in most things teenagers were interested in. I constantly had girls interested in hooking up, but I had no time for distractions.

I remember one girl who lived in Greensborough, a suburb to the north-east of Melbourne. She and I would talk over the phone, and I would try to get her off as quickly as possible so I could get back to business.

She would call to invite me to her place so we could swim in her pool, openly declaring that her parents were away, and she was home alone.

The commute to her place would have taken 45 minutes each way, which equated to 3 hours of lost productivity in my warped mind.

We went out on a few dates to the city to watch a few movies, only because my mate had been receiving calls from her friend trying to arrange a double-date.

She was very keen, but I was obsessed with computers at the time and had no room in my life for a girlfriend. And I had too much respect for her to string her along. So, I would ghost her; not a good thing to do.

Don't get me wrong, she was beautiful. A great personality and a smile that lit up a room. She had long flowing chestnut hair and a slender athletic physique which would stop traffic. The perfect girlfriend for any teenage boy with half a brain. Marriage material.

But my interests were elsewhere.

There were others of course. Casual encounters. Ships passing through the night. Nothing serious though.

When you train your entire life to be the best and you reach the pinnacle, the last thing you want to do is drop the ball!

Being in the zone is a strange feeling. You have the respect and admiration of the entire community. You are well known by your peers locally and internationally. You are at the top of the international pirating and hacking ecosystem.

Power. Privilege. Respect.

For an adolescent teenager, these feelings fuel you to go further. To become even better. To take more risks. To achieve greater things.

Many don't understand why elite individuals are so dedicated and committed to their chosen field. They can't see why it's so important. Athletes do it for the money. Scholars do it for prestige. Doctors do it to save lives. So, what purpose were hackers, crackers, and pirates doing all this work. There was never any pot of gold at the end of the rainbow. No large sponsorship deals around the corner. No hot lingerie models lining up to date you.

The truth was that we did it to pump up our own fragile egos. Living in a world which shunned misfits, the online computer world was somewhere special. A place where we could become masters of the universe. *Our* universe.

The vast majority of people in the community were academically strong. They excelled in their studies (if they chose to apply themselves). Yet they felt trapped in their environment. Following the bouncing ball from year to year. Listening to their parents tell them they needed to become doctors or lawyers to prove their worth.

"Finish primary school, graduate from high school, graduate from university, get married, have kids, buy a house, work hard for 35 years, then retire."

This is the roadmap to success, right?

Wrong!

This is what parents want their children to do, not necessarily what should be done. Everyone will carve out their own path. There is a beginning and an end. What happens in between is what we make of it. Success is not measured in one thing, it's the culmination of everything.

Ultimately, I believe you need to do what makes you happy, not what makes others happy.

Computers started as a hobby for most and ended up becoming an obsession. The transition was gradual. One day you're sitting in a computer lab playing some games with friends. The next, you're downloading thousands of credit card numbers from a private London BBS and uploading them to a site in California.

It's like the boiling frog experiment we did in high school science. Most science students would know the experiment.

If you place a frog in a pot of boiling water, it will immediately jump out. If, however, you placed the frog in a pot with room temperature water, then put it on the burner and gradually heated the water, the frog would happily sit in the pot even though the water temperature might exceed the first experiment.

We were the frog. We didn't jump straight into hot water; we wouldn't have done that. We started out innocently and transitioned into what we became, never considering the changes happening around us.

The more connected I was online, the less connected I was in the real world. I had descended into a world which was exciting, new, innovative. It wasn't real, but I thought it was.

Every addiction starts off slowly. A little taste here, a weekend binge there. Before you know it, you are trapped in a world you once thought you controlled, but now it was controlling you. You can't get out. You can't stop. Your mind and body crave the rush too much. Everything you are is because of this addiction. You're in the zone, and nothing else matters. The real world becomes foggy. A distraction. Somewhere you don't want to live. Only your newfound reality is what matters.

The water was boiling, but I was happy to sit in the pot.

How to Hack and Crack - Retro Style

When you're in the underground, you need to have a pseudonym. I needed two names. Anything that I did on the software pirating or the cracking side, I would use the name 'Cobra Strike'.

Whenever I downloaded games or pieces of software from overseas, many would have some pre-roll promotions from the people that cracked that particular copy. Because these games were coming through my hands as a distribution point, I would use specific tools at my disposal to break into the cracker's pre-roll code and make changes by adding my own branding and unique commentary.

When it came to anything online-related, I went by the name "The Rebel", which was what I used both online and when attending the BBS catchups in the Melbourne CBD.

I kept the two separate, and always made sure that they never crossed over. I wanted to ensure people knew me as either one or the other, but not the same person.

My reasoning was straight-forward. In the pirating community, it was your opportunity to promote yourself to a large audience. Every person who received the software you distributed would know it came from you. Soon, your name was plastered all over the place.

With online activities, you wanted to stay anonymous. It wasn't about promotion; it was more about the technology and growing your knowledge and capabilities. The more you learned, the higher you got in the hierarchy, the less you spoke. The top guys would never discuss what they were up to. Only amongst each other. There was a deep circle of trust which was binding. People who blabbed were quickly expelled from the inner sanctum.

The main role of a cracker was to bypass the copy protection located on a licensed floppy disk, so that software could be copied multiple times. To do this, the cracker needed to have a copy of the original protected floppy disk.

Copy protection was typically stored at the exact location of one or many tracks and sectors on the original disk. When a disk is formatted at factory and the software installed on it, each track and sector is numbered. Therefore, as an example, the protection key may be stored on Track 20 Sector 12. Or it could be stored across multiple tracks and multiple sectors.

When the original software fired up, it would instruct the disk drive to go and check the specific track and sector to make sure that that particular code was there, and it was valid. If the code was not there, or it was not valid, the software would not fire up. The drive would continue spinning in a "null" state; the software basically instructed the disk to continue spinning aimlessly.

There were two methods used to crack a piece of licensed software.

One way was to make an exact copy of the disk, including the numbering sequence of that specific disk. This way, the software on the disk was copied to the exact location as the original disk. The numbering of the tracks and sectors would be an exact match. You only needed to make sure the protection key was saved to the exact track and sector needed to ensure the copied software fired up.

Sometimes the copy protection would run across multiple sectors and across half sectors. You might therefore have a situation where the first half of a sector might be blank, and the second half contained part of the protection key. Sophisticated protection keys were split across multiple tracks and sectors across the disk, making it even more tricky to copy.

Keys might spill into the next sector which had another couple of characters, or across other sectors of the disk. When you combined all the characters across the sectors in the correct sequence, it revealed the protection key. This was the code that the copy protection was looking for to unlock the software.

To get around this, you would figure out exactly what tracks and sectors contained the specific copy protection codes. The only way to do this was to use a hexadecimal converter to view the code, then search for the copy protection calls which were usually found in the first few sections of the software. Once you discovered this, you could then work out where the protection codes were stored on a disk.

You would then run a very sophisticated copying tool that would not just do an exact image of the actual disk but will ensure that the exact track and sector that held this protection mechanism would be stored in the same place as the disk that you were making the copy on.

For example, if the protection was on Track 20 Sector 2, the disk you were copying must have the protection code on Track 20 Sector 2, the exact spot. It really didn't matter where on the disk the software was running, but that protection key had to be on that very specific track and sector.

My favorite tool of choice at the time was called 'Marauder', which allowed you to punch in tracks and sectors where you wanted to get exact copies of that track and sector.

You would effectively build an instruction set which Marauder would read, and then follow the sequence of which tracks and sectors to copy and in which order. Once you perfected the script, you would save it on the disk and then distribute the copied version to others.

If they wanted to make a copy, they would simply load up the script into their version of Marauder, insert the disk, and Marauder would follow the script precisely as instructed, which would make an exact copy of the licensed software.

Now, this was a very cumbersome process. You needed to have a deep understanding of how floppy disks worked, how computer programs were written, be able to scan disks looking for specific code, look for areas where the software was pointing to the protection key, how many pieces of a key were on a specific disk etc.

Then, you needed to check the compiled code to make sure it matched what the software was looking for. After that, you needed to create the Marauder script, test it (which almost never worked the first time), and spend time deciphering what you were missing.

It was a very time-consuming process!

In addition, it was difficult to distribute disks at scale if you did not have the Marauder software or similar software to be able to do exact track and sector copies. For a novice wanting to simply make a copy of the software, if they didn't have the right tools or understanding of how it worked, there was no way it would work for them.

People would come to me and say, "I tried copying this, but it didn't work". They obviously lacked the deep knowledge and understanding of what they were doing.

"Uber lamer!", I would think.

Most people tried to avoid this method, because unless you knew exactly what you were doing, it was pointless.

The second method, which was much better in many ways, was where you would crack the copy protection mechanism entirely.

Using hexadecimal tools, you would search the software for the specific code which requests jumping to check the security protection on the disk and returns the result of this check.

You could do one of several things. You could change the code with your hexadecimal editor so that when the code did the protection check, the code would always say that the key was valid. Effectively, you were changing the code to hard-wire the check to always say that it was valid, regardless of what the result of the check was.

The other way was to simply bypass the entire protection check process altogether, by changing the code so that it would skip the protection check and go straight into the game itself.

Regardless of which method you used, from a user's perspective, the result was that the copy protection was nullified, and the game would load seamlessly. This was a much better method for copying software, as it didn't require physical disks to be exchanged and could therefore be shared as a file from person to person, especially online.

This method is still being used today to circumvent licensed software. Whilst the delivery method has changed, the underlying methodology remains the same. Instead of checking a specific track and sector, the software checks the online server to verify an active account.

Hacking on the other hand is very, very different. Hacking is all about breaking into a computer system, and that required a very different level of skill.

Effectively, you would first need to identify the access point that you would use to get into the system. This would either be a direct dial-in number, or a link from another computer.

For example, in the late 1980's many of the universities and colleges had interconnected their computers for information sharing purposes. The theory at the time was that if someone was on a particular system, they were a verified and validated user who was allowed to roam around the network.

Most systems administrators were trailblazers who were more interested in connecting systems together to increase reach for research purposes. They failed to understand or appreciate the underground hacking community who were looking for ways to get on one system, then jump around to other systems.

In fact, university and college systems were some of the least protected systems you could access. And they allowed you to freely roam across other systems with relative ease.

Once you were able to access a system, you then needed a login and password, preferably at the "administrator" level, which would give you access to the complete system. Without administrator access, you were restricted in where you could go and what folders and, more importantly, what files you were able to access.

There were several ways to get the login credentials. The first way was to run a brute force software program which had a dictionary library sitting behind it. The software would try to login as an administrator. The login screen was very simple: a login name and password prompt would be displayed along with the usual prohibited access warnings. The program would always default to 'admin' as username, and then cycle through the dictionary file.

One of the most important tasks as a hacker was to ensure your dictionary file was well balanced and constantly refreshed. You had to balance the file size to contain words which were most likely to be used by an administrator. For example, most systems had a minimum character restriction, so you would remove anything with 4 letters or less. Additionally, most administrators would not have a password length greater than 8 or 9 characters, as this was too difficult to remember. So, you would tune your dictionary file to add as many words as possible, but at the same time keeping the file size small enough so that it didn't spend weeks rotating through the file.

Hackers would spend a lot of time tuning their dictionary files. These proved to be a very hot commodity which was hardly ever shared within the community. Sure, you could get access to a dictionary file, but it may not have been properly tuned by an experienced, highly skilled, successful hacker.

The dictionary file could have thousands of potential password combinations, and the program would constantly rotate through each and every word from A to Z. Hence why it was referred to as a brute force attack.

Gradually, hackers realized that running a brute force attack from your home computer wasn't ideal. Firstly, you had limited computing power to run the attack, meaning the program would take a long time to cycle through the dictionary. Tying up the phone line for days wasn't an option. Secondly, there was an inherent risk of the line being traced back to your home phone account which meant a possible visit from the authorities.

The solution was simple yet effective. By spreading the brute force attack across multiple computers, you were able to significantly increase your computing power and therefore reduce the time to run through your dictionary. University computers were the best place to run your program. Hence why university and college accounts were valuable to a hacker.

Today, the same methodology is used by hackers to run both brute force attacks, as well as Distributed Denial of Service attacks (known as 'DDoS attacks'). These DDoS attacks effectively bring down a system through excessive and relentless multiple systems (or 'bots') attacking simultaneously.

The concept of a distributed server attack was developed in the 1980's by the global computer underground and is still in use today.

This approach was tried and tested and was a standard tool that any credible hacker would use.

In the 1980's, many administrators who set up servers would make two fatal mistakes.

When an administrator first set up server software, the default login administrator password out-of-the-box was 'password'. Often, administrators simply would not change the administrator password as they either got caught up doing other things or didn't think it was a big deal. After all, they were the only ones who had access to the system terminal, and therefore didn't understand or appreciate the security risk they had left open.

Once an administrator installs the new server, the very first thing they need to do is change the administrator password to something else. The next thing they should do is implement a rotation method where they would regularly change their administrator password. It should be changed at least every two months as a minimum.

Again, back in the 1980's, a lot of administrators did not change the admin password.

This was widespread knowledge in the hacker community. In fact, it's probably the first thing you learn when you aspire to become a hacker. Back then, it was surprising how many system administrators just used the default password, and never changed it.

The other thing they failed to change, which sometimes was a default setting, was to allow for unlimited login attempts. The common practice these days is three attempts, then a lock-out and notification sent to the administrator.

Often administrators would use the same password across multiple servers; A massive no-no! You might as well leave the keys under the door mat.

Back then, a lot of system administrators did not even think that some teenage hacker sitting at home in their underwear was going to get onto their system somehow and run a brute force software program with their dictionary file to get through the system.

A brute force hack was probably the standard tool that a hacker would use. But there were two other ways that hackers would get into systems, though they were a lot more difficult, but definitely possible.

The first way is a physical security breach. This is where a hacker would somehow gain access to the actual terminal of a system. For example, you may be a university student working in a computer lab, or you are at a university campus and you have somehow gotten access to the computer room for whatever reason. From the terminal, you can access the actual system. And if the terminal screen was unlocked and the administrator account was active, you could go in and set up a new administrator login and password without anyone knowing.

You would then dial in later from home, or in the computer lab of the campus, and access the system as an administrator.

This method was not common because you needed to be at a particular computer which had administrator account access, and no-one around. However, this was a way to get on without any detection.

Once you gained access to one university or college system, you effectively had access to other systems connected to the network. At the time, university and college computer systems were being connected to each other for research purposes.

You might have a particular system in Melbourne that connected to a system in Sydney, for example, for education purposes. If you could therefore get the login and password on the Melbourne system, you could then automatically start connecting to other systems on the network. This was effectively a gateway method to get an account locally that would then open other doors into other systems that are connected to each other.

The second way also needed more sophistication and physical access. You would run a network sniffer program on a live network and capture login credentials. This involved getting onto a target network computer, running a piece of software which would sit on the actual network and look for login name and passwords. Typically, a lot of these internal network communications were not encrypted. The assumption was that if you are physically on the network, then you were a trusted source.

All you had to do was type in the keywords you were looking for, such as "admin", and wait for the software to capture data about that keyword. Any time the word "admin" was transmitted across the internal network, you would copy the packets that were being transmitted. You would then open the packets on your software and bingo: admin password.

Once you had access to the administrator password, you would log in remotely whilst on the network, and create a new administrator account with your own password (otherwise known as a 'back door').

The trick was that you did not use the admin account, because if they followed proper protocol, they would be updating their passwords every one or two months. The goal was to log in as an administrator, create a new replica administrator account, use a unique 'normal' name, with your own set of password credentials that you controlled.

You did not call it an admin account. You would check the user file to review the standard naming conventions for that particular system, and make sure that the name you created matched the convention. This way, if anyone was flicking through the user logs to try and figure out who was who, it would blend into all the other names that were there. It was less likely that this particular account would be discovered if they looked properly.

Normally when you search for users, the system would list which users are standard and which users have administrator permissions, but many times this was overlooked by the administrators.

When you are talking about large corporations or campuses which have multiple system administrators that rotate through, it is not uncommon for one not to know what the other one is doing, or to be unaware of how many users or admin accounts there are. It is common for these things to be overlooked.

Then there's what I like to refer to as "Old Faithful". A sure-fire way to get a login name and password.

Basically, you would log on to a terminal in your computer lab using your own user credentials. You would then run a program you had written which would recreate an exact copy of the official terminal login screen.

You would then walk away, stand nearby, and wait.

When another user sits down and enters their username and password, your program will capture their credentials and store them in your account, and then display "Incorrect Username or Password" on the screen. Your program would then automatically log out from your account and restart the terminal with the official login screen. The user will think they inadvertently entered the wrong details and carefully re-enter them again.

This method was a good way to capture student login credentials, but occasionally, if you were lucky, an administrator would try logging in via your compromised terminal, resulting in the highly prized admin credentials being captured.

Like a Venus Flytrap, once they typed in their details and pressed ENTER, they had sealed their fate.

Pecking Order

There were generally five different types of people in the global computer underground.

Pirates, which was the area I fell into, specialized in distributing software and games in high volumes to the public. Hackers, who specialized in breaking into connected computer systems both locally and abroad. Crackers, who specialized in breaking copy protection used on licensed floppy disk to replicate them for mass distribution. Phreakers, who specialized in getting free phone calls. And Phishers, who specialized in getting personal information from the public for their own benefit, such as obtaining credit card numbers and calling cards.

There was therefore a clear pecking order. At the top of the pyramid were the Hackers, followed by the Crackers, then Pirates, then Phreakers, and finally Phishers.

Some, like me, would cross over a number of categories depending on what I was doing at the time. Others would start out in one category and then transition to another. Most would identify themselves in one category.

Phreakers would do things like run a physical telephone wire from an actual telecommunications junction box located at a street corner or from a neighbor's telephone line. Some would run copper wires directly to their house so that they could use someone else's line to make free calls. Some would have mobile units, driving around in their car searching for a quiet location out of the way. They would have their computers wired up to batteries, and a long copper cable which they would run along the street to the junction box. Very unsophisticated, high-risk activities, which really didn't appeal to me.

In the US, some phreakers were using the legendary Blue Box device which was made famous by the Apple founders Steve Jobs and Steve Wozniak. These devices allowed the user to make free long-distance calls from a public telephone booth. The system would only work in the US as their carrier network identified the signals being transmitted from the device.

Whilst a Blue Box was an interesting piece of kit, it wasn't really that valuable to a hacker or someone online. You see, people accessing the BBSs needed to use their computers with a modem to access other computers. In the 1980's, it was virtually impossible to lug your home computer and modem to a local phone box, set up, and jump online. Laptop computers were something you saw in the movies. They didn't become available until later and were very expensive when they were first released.

Plus, the Blue Box devices started becoming obsolete in the late 1980's when the carrier systems were upgraded to circumvent their signal transmissions.

It was, however, great if you simply wanted to make free long-distance voice phone calls.

Phishing is very prominent today. I'm certain you will have received an email from someone asking you to click on a link, or to provide some information to verify something with a provider. Phishing activities have exploded since the dawn of the Internet. However, these activities were going on in the 1980's, if not earlier, just far less sophisticated.

Back in the 1980's, some of you may remember that credit card payments used to be taken at restaurants and stores by using a mechanical machine and carbon or ink paper. The merchant would place your credit card onto a mechanical device, and then place a credit card docket which contained three layered pieces of paper on top of the card. In between each layer, there would be a piece of carbon or ink paper. The merchant would then physically swipe the mechanical device forward and backward across the face of the credit card. This process would then imprint the raised numbers and details of the actual credit card onto the three paper copies and, more importantly, on the actual carbon or ink paper.

Three copies were created: one for the merchant, one for the customer, and one for the credit card company. The carbon or ink paper in the middle, which was used to transfer the information from the card to the papers, was then discarded. Merchants would usually screw up the carbon or ink paper, and throw it in the bin.

What they should have been doing is tearing up that slip so that it was properly destroyed. The reason: that carbon or ink paper slip contained an exact imprint of a credit card with full customer details. Merchants would usually discard these carbon or ink papers in their dumpster, right out the back of their store without even thinking about it.

Phishers would visit various merchants such as restaurants, bars, and retail stores, in the early hours of the morning. They would then jump in the dumpsters, specifically searching for these carbon or ink paper copies of credit cards. Once they found what they were looking for, they would capture the credit card details and obtain valid credit cards.

This practice was considered very low in the pecking order, hence why Phishing was a disrespected activity. You'd have to be a certain type of person to jump in a dumpster out the back of a restaurant at 3:00am, to find these papers amongst the discarded food and other rubbish. The smell alone would turn most people off this activity.

I could never understand why anyone would want to do this as a hobby. To me, they simply lacked the skills or discipline to source this information online. They were trained chimps with low IQs looking to get their kicks on a slow night out on the town. Scratching through garbage hoping to strike gold. More often, they would hit a bad Beef Vindaloo or half-eaten Spaghetti Bolognese, I thought.

But if you wanted valid credit card numbers, that was one of the easiest ways to get them.

I imagined these people competing with the rats of the laneways, fighting for scraps left behind by diners after a night out celebrating Pop's 85th birthday. To me, I couldn't distinguish between the street rats or the rodents living in the sewers.

Phishers would identify a pattern in their dumpster jumps. They would get a 'feel' for which specific merchants were the most careless when it came to discarding the credit card carbon or ink papers, and which ones disposed of them correctly. They would therefore know which merchants to target.

But online, Phishers bragged about their bin jumping exploits. They saw it as an adventure. They enjoyed the thrill of roaming the streets looking for those elusive credit card numbers. They would brag about their exploits, outline which types of business yielded the best results, and what areas they had "hit".

The credit cards were often offered as barter for other items of value on the BBSs like calling cards and pirated games.

Thankfully, I rarely visited the sub-basement of Zen and PI to read about their exploits. I had a peek into their world through their forums, but never engaged with them.

There was always a battle between hackers and pirates. Pirates always believed they were just collectors. They weren't doing any damage to anyone. All they were doing was basically sharing files. There was no harm in the discipline, as long as we didn't charge for software. It was a hobby for us which was born out of necessity: we didn't have the funds to buy software.

Hackers viewed pirates as being beneath them because they believed it was easy to copy software. There was nothing intellectual about it in their minds. Whereas being able to hack into a system required discipline, a certain amount of knowledge, and certain specific tools. Hacking was more involved, more intricate.

Given that I had skills in both fields, albeit at a basic hacking level, there was a lot more involved in hacking than there was in being a basic pirate. I personally didn't enjoy hacking. I didn't get the same adrenaline rush as I did when I was cracking or pirating.

Hackers genuinely believed they were at the top of the tree. They were the people who spent a lot of time and energy breaking into computers. They needed a high intellect to do what they did.

You needed to have a basic understanding of how telephony worked, as well as the operating systems you were accessing. The most popular ones at the time were VMS and Unix, and they each had their own unique idiosyncrasies. You needed to know how to write programs which would help you automate some of the mundane functions a hacker would do, such as demon dialers and brute force password attack programs. You also required a broad understanding of computers and network infrastructure. A lot to learn for a teenager.

Hacking required incredible discipline and a lot of time. So why do it? Well, there are a few reasons.

Hackers enjoyed the thrill of the hunt; breaking into a system was something elusive. When a new system was discovered, it was like the new hot chick in school. Guys would flock to her to see who was good enough to win her phone number. They would spend hours trying to break into a system, to unlock the secrets hiding behind the login screen.

Every hacker knows the euphoria when they finally succeed in breaking into a system. It's like a drug user shooting a hit straight into their arm and experiencing a high that can't be explained to non-users. Pure unbridled ecstasy.

But getting in was only half the battle. To keep the adrenaline pumping, a hacker would then snoop around to see what treasures they had uncovered. No point breaking into a temple if it was empty. Where's the loot, the payoff, the shiny thing that made all the effort worthwhile.

In most cases, the payoff was information. Files tucked away where nobody was meant to find them. Emails stored on a mail server. Sensitive reports reserved for senior management.

And depending on the system, the information value would greatly change. A university computer network would yield storage and processing power. Somewhere where you could store files you've lifted from other systems. A rat hole that only you knew existed. Somewhere where you could store confidential, often incriminating evidence. Like storing food in a fridge for later when you get hungry.

Commercial and military systems were a completely different story. These systems contained valuable information about all sorts of top-secret projects and activities.

Some commercial systems would contain documents and email exchanges about corporate strategies or merger discussions; super valuable for anyone looking for market sensitive information that could move a share price.

Then there were the military systems, which were known to have documents outlining newly developed weapons and equipment. Missile blueprints and specifications, purchase orders from government defense forces, tender documents. All sorts of valuable stuff.

The golden rule for any credible hacker was to ensure you left the place clean and tidy when you left. Don't damage anything which would raise suspicion that you had entered. Be like a cat; nimble, quiet, quick. Get in, snoop around, map the system, copy what you wanted, and get out. The last thing you wanted to do was draw attention to yourself or your fellow hackers.

The top hackers all followed a simple code of conduct: never tell, never brag, always stay in the shadows. The best hackers in the world were the people that kept their mouths shut. The inner sanctum of the hacker community was built on extreme trust and silence.

Most of them would never post anything on the public forums, and very little on the private sections of the BBS. If you were in the public area of Zen and a hacker was on at the same time, and you reached out to them for a chat, they would completely ghost you. They weren't there to make friends and talk about their school day. They were there to access information from other hackers on a project they were working on, or a target they had identified, or sharing access codes to a computer they had hacked into.

Typically, they would talk only amongst themselves. They wouldn't talk to most of the people on the private section either. Remember, in the hacker's mind, they were the 800-pound gorilla and everyone else was beneath them. They had this sense of power and privilege, and they weren't going to mingle with the riff raff. Especially not with pirates, who they deemed peasants.

Hackers would always have rat holes all over the systems they infiltrated. Storage areas where they could hide their information and access it later. Most of the time, they would not have anything stored on their local computer. It was too dangerous. If there was ever a raid, they would be caught red-handed.

However, many didn't care about the risk of getting caught. There was a genuine belief that they were untouchables; too smart for the authorities who were still using typewriters to write up reports.

Hackers were therefore at the top of the tree and were the most revered people. Even pop culture portrays hackers as God-like people with incredible powers who could cause havoc and mayhem with a click of their fingers.

The Hollywood stereotype couldn't be further from reality.

At the time, the best hacking group in Australia, and probably one of the best in the world, were a group called "The Realm". This group, rumored to be made up of five hackers, were the most mysterious group in the inner sanctum. They were very disciplined in what they did. They were extremely secretive and very exclusive. They would never chat with anyone on either the public or private forums. The only way to chat with these people was if you were vetted. They needed to know who you were, that you were reliable, and you shared the same discipline to secrecy as they did. In short, you had to be trustworthy.

I was very fortunate to have some communication with a few members of The Realm. The only reason was because I was networking with others on the private BBS sections and attended the regular catchups in the Melbourne CBD. I never actually physically met any members of the group, but I suspect a few of them were also attending the Melbourne CBD catchups and were sizing me up.

The Realm were way ahead of any other hackers that I chatted with. Everything they did was very secretive and shrouded in urban myth.

Whenever there was a hack of any kind announced on the TV or in the papers, the BBS systems would light up with rumors about the exploits of this mythical hacker team. They were linked to every sophisticated and brazen hack that was made public in the 1980's, even the hacks that didn't make the six-o'clock news or mainstream newspapers.

The members of The Realm wouldn't talk about what they did, only amongst themselves. That was the discipline of a top hacker: never speak about anything you do.

I had managed to build myself up to a level where I was in the inner sanctum of BBSs in Germany, Holland, UK, and the US to name a few. I was also in the inner sanctum of some of the top hackers in the country. As a 16-year-old, it was an amazing experience.

I learned a lot about telephony, infrastructure, networks, servers, operating systems, password protection, and cyber security. I always wanted to have a very deep understanding of how every single part of a system worked and how the parts interconnected. I was especially curious as to how it could be exploited, and where the gaps were.

Like Neo fighting off the agents in the final scene of the movie 'The Matrix', my mind worked very differently to everyone else. Where others saw the physical layer, I would see the digital layers that sat behind the facade. I would map out entire systems in my mind whereby I identified each component, creating a blueprint for how something really worked. Once you get to this point, you start seeing the digital world for what it is; a series of interconnected components that harmoniously coexist.

That thirst for knowledge and understanding has stayed with me to this very day. It's in my DNA.

Hackers used to look at systems from a network level, how to identify systems and break into them. They also needed to understand the operating system in terms of knowing how to get through the login and password protection to be able to create administrator accounts, which gave them unlimited access rights to the systems. They would then need to be able to sniff out other systems and vulnerabilities on the system, what other systems they can connect to, and where they can jump to.

It is a highly sophisticated endeavor, and you need a lot of discipline.

Personally, my skills transcended hacking, cracking, and pirating. I was a coder and an infrastructure guy. I always believed that having skills in all three disciplines was essential to building a solid foundation for computers. I also believed that having a blend of all three disciplines was far more important than specializing in just one.

Hackers might have thought they were at the top of the tree, but for me, people who possessed these three disciplines were at the top of the mountain.

Myths and Legends

There were always numerous hacks discussed on the private forums of the BBS systems which had become folklore. Whether they're true or not, I cannot say. But there were some great yarns shared on the BBSs at the time.

One hack that I thought was amazing related to a very large publicly traded company in Australia. The hackers managed to access confidential Board Papers and private senior management email exchanges, which outlined details of a soon to be executed acquisition of another publicly traded company. The Board had agreed to submit a bid to buy the target company for a specific price, along with the timing of when the buy-out would occur.

Typically, when a company makes a formal takeover offer to acquire another publicly traded company, the price of the target company goes up proportionate to the offer made per share. In most cases, the acquiring company's share price drops suddenly as soon as the takeover offer is announced.

Having access to this inside information, which of course is highly illegal, the hackers then accessed the financial systems of this particular company and transferred funds out of that bank account into another bank account. They then used the other bank account to purchase shares in the target company before the announcement was made to the public.

As soon as the company announcement was made public and the target company shares rose sharply, the hackers sold all the shares they had purchased with the acquiring company's money. They then transferred the original money that they'd taken back into the publicly listed company accounts.

From a books and records perspective, it just seemed like it was a normal transaction: money in and money out. No money was missing, and it therefore got through the auditors.

The hackers had used the company's money to arbitrage a transaction using information obtained from the Board Papers and emails they had accessed. From what I understand, and I can't verify whether it's true or not, they got away with it.

The second hack, not so smart.

A hacker had managed to access the ordering system for one of Australia's largest automakers and was playing around. He decided, rather cheekily, to order seven brand new top-of-the-range vehicles from this automaker.

He placed the order through the system, uniquely pimping out each vehicle with custom factory extras. Leather seats, larger wheels, tinted windows, sunroof. Black paintwork with red trim for one, Midnight blue paintwork for another.

"Why not?", he thought as he filled out the online purchase order.

Strangely, everything got validated, and the order confirmed. He thought the system would pick up the error and cancel the order. But it didn't. A delivery date was confirmed, and a printout of the order confirmation was made available. Everything looked as it should.

The delivery day came, and curiosity kicked in. He decided to go to the dealership where the cars were going to be delivered to see if the order went through.

"Surely it wouldn't be that easy!", he thought.

The hacker, who happened to be 19 years old, turned up to the dealership in his beat-up hatchback and asked if the vehicles he had ordered had been delivered. The sales representative checked the system, grabbed the paperwork, and escorted the hacker to the yard.

To his surprise and disbelief, he stood in front of seven brand new top of the range vehicles, which were built to the exact specifications he had ordered through the system.

The paperwork said the vehicles had been paid for and were to be collected by the customer.

Unfortunately for the hacker, things went south from there.

The dealership took one look at the hacker and thought: "Hang on, this 19-year-old kid has turned up to collect seven top-of-the-range vehicles. He's come by himself in a beat-up shit heap. How's he going to drive seven cars off the lot?". Very suspect.

After double-checking the order to figure out what had happened, in the end the hacker got caught.

Dumb move.

A Pirate's Life

Logging on to Zen and Pacific Island during peak times proved to be difficult, particularly from around 5:00 PM up to about 11:00 PM. People would try to log in after school and after dinner. Given their growing popularity, both Pacific Island and Zen had time restrictions. Normally a new user would be allocated 30 minutes duration. Once your time expired, you were automatically kicked off.

Regular users would have up to 60 minutes, but they would need to be well known by the SysOp, or contribute something to the community such as hardware, software, or pay a 'donation'.

Craig Bowen ran both Zen and PI from his bedroom. He lived in the south-eastern suburbs of Melbourne, somewhere between Glen Waverley and Clayton. I never personally visited his home, but others in the community had been there. They saw first-hand the two BBSs sitting side by side humming away.

Craig had five phone lines coming into his house. One for PI, and four for Zen. Each line would cost roughly $40 per month in line rental charges from Telecom Australia. That's a cool $200 per month in rental alone, plus any other calls he would make for his own hacking exploits. In addition, these systems ran 24/7/365. Never down, never offline. Craig prided himself on having very reliable systems.

At one point, Craig needed to buy a new hard drive for Zen. The volume of files being shared on the system was reaching capacity, and the system was at a critical point. Craig sent out a request to the community for 'donations' to buy a new hard drive which would significantly increase capacity and improve performance.

Sure enough, the community responded. Letters trickled in containing $5 and $10 bills. High school students like me sent through their lunch money to ensure Craig could get what he needed to keep the music playing.

We needed to keep feeding our addiction.

The investment in such a hobby would have been significant. Each server would cost roughly $2,500 to purchase, not to mention the modem costs which easily set you back $500 a-piece back then. Add other bits and pieces like rotary dialers and power surge protectors, and you wouldn't get much change for a cool $10 grand. Throw in the running costs, electricity bills, telephony, and Craig's sweat equity, and you quickly realize that BBSs were an expensive hobby to operate.

So why would anyone want to do it?

You needed to be in the community to understand the answer. Owning and operating a BBS meant you had street cred. Owning the top two BBSs in Australia gave you incredible kudos. Owning two of the top BBSs in the world, well, you're now in rarified air.

Was it worth it? Absolutely!

As a member of the private section, uploading games and software meant that you needed a lot more time than one hour. The SysOp would ratchet up your allocation time to two hours, which gave you plenty of time to upload the games that you were pirating, as well as download whatever material you wanted. Of course, when your two hours were up and you got kicked off, you immediately dialed back in to start another two-hour session.

Spending so much time on the BBS systems, particularly on Zen, you bumped into a lot of people. Zen had a four-line rotary system, so up to four people could be logged on and active at once.

When you first log onto Zen, the first thing you would do is see who else was logged on. If you knew other users who were active at the same time as you, you'd quickly invite them into either a public or private chat forum. That was one of the best ways to meet the people in the community.

My typical login time was usually around 9:00pm, and I'd typically be online until about midnight, but often to 2:00am. If I was doing a big pirate session when I was uploading a lot of software that I'd downloaded from overseas, I'd be on until 4:00am or 5:00am in the morning, have a short sleep, up at 7:30am and off to school.

This made me one of the biggest users of Zen, because I needed the time to be able to upload all the software I was getting from overseas.

But there was one guy who was almost as active as I was. He wasn't a pirate, or anything like that. He was a computer enthusiast. Almost every time I logged on to Zen, this person would reach out to me and ask me to join him in a private chat. It turns out this guy was two years younger than me; he would have been 14 years old at the time. This person lived in Coldstream, a rural town just outside the Melbourne metropolitan area.

This was important at the time. Back in the 1980's, if you lived within the Melbourne metropolitan area, every phone call within the area was charged as a single flat fee of $0.20 per call. Regardless of how long you stayed on the phone, as long as the call was a local call, it was $0.20 per call.

However, if you lived outside the Melbourne metropolitan area, you had to pay what were referred to as STD rates. These rates were timed rates, which equated to around $0.20 per minute. The longer you stayed on the call, the more you paid.

STD calls were usually reserved for calling interstate, such as between Melbourne and Sydney. However, the telecommunications provider at the time used to charge STD rates for all calls made outside a local area. For example, if you lived in Ballarat, which is a town 1.5 hours outside of Melbourne, and you called within the local zone of Ballarat, you would pay $0.20 flat rate for the call. However, if you called Melbourne from Ballarat, you would be charged STD rates which were timed calls. Rates varied between towns and cities, so you needed to understand how the phone system worked so that you weren't racking up exorbitant fees.

Because of this, most in the BBS community (and the general community for that matter) despised Telecom Australia. They were seen as a bloated monopoly who were constantly ripping off their customers. Whenever Telecom was mentioned in any conversation, oral or written, a barrage of insults would follow. Some referred to them as 'TeleScum'.

Anyway, this kid was on Zen every night. He was on there a lot. Every time I logged on, before I could even start typing, I would receive a chat request from this kid. He was almost like the local pest!

He kept messaging me, saying things like "How are you Rebel?", "Let's have a chat, come to Chat 111", "What are you doing?", etc.

It seemed like he was really bored; he kept wanting to talk to me. I was doing my own thing. I was busy uploading games, I was chatting with other pirates in the private forums, I was chatting to hackers. The last thing I wanted to do was to sit there talking to some young kid from Coldstream.

Now, at the time it was all about the pecking order. When I first started out with my one floppy disk and my one game, I was basically the same as the pest from Coldstream. I would sit there pestering everyone to try and get as much information as possible.

I therefore understood where he was coming from. He was a kid trying to learn the ropes.

So, every now and then, I'd chat. He'd asked me all sorts of questions like "How are you?", "What's happening?", "Have you heard of this?", "Have you heard that?", "Oh, I hear there's a private section of the board, do you know anything about that?"

"Man, this guy sure likes to gossip!", I thought.

Of course, you deny everything because as part of the inner sanctum, you never spoke about this stuff to anyone outside the group. Failure to follow the rules meant you would be expelled. Reputation was supremely important. Information was the currency which separated the haves from the have nots.

I got to meet a lot of different people on the BBS systems and at the meets.

Ultimately, it was all about where you sat in the pecking order. The higher up the ladder you got, the more they wanted to get to know you. Being active on the BBSs was a sure-fire way to attract attention from others loitering around looking for someone to chat with. It was, after all, a community where people were trying to get to know each other and learn about technology.

When it came to the Pest, we just happened to be at different levels at that point.

When you're at the top, you avoided socializing with too many people. There was a trust factor; if you didn't know the person, or they weren't referred to you and vouched for by someone you trusted in the group, you wouldn't speak to them. You would be civil, and conduct general chit chat, but you definitely wouldn't talk about what you were up to. That was an absolute no-no.

Another regular who I did spend time chatting with was 'Masked Avenger'. He was one year younger than me but was always funny on the forums. He and I had met a few times at the Melbourne CBD catchups, so I knew he was part of the inner sanctum.

'Avenger', as most referred to him at the time, was a little high-strung. A nice guy overall, but someone who seemed to always be on the edge. He reminded me of one of the characters in 'The Matrix'; the one who designed the red lady in the simulator. Brilliant with what he could do, but just a little out there.

Avenger was a regular, and someone who could be trusted. To a certain level. Whilst he was more on the pirating side (hence why we both got along so well), I never divulged anything to do with the hacker discussions I was having. Remember, you needed the discipline to separate the two worlds.

Radioactive eventually bought a modem and had found his way onto Zen and PI. He would reach out to me with the same rumors about the private sections. I had known him for many years prior and respected him as a person.

But I was now at a different level to him. I had been on the boards for many months before he turned up. I was more than happy to supply him with the latest and greatest games, as I trusted him more than most.

"I hear there's a private section of Zen and PI. Do you know anything about it?", he would ask.

I know Radioactive was trustworthy, so if push came to shove and someone asked me to vouch for him, I would have done it in a heartbeat. But the private section was reserved for the elite, and whilst Radioactive was certainly more than capable technically, he probably wasn't someone who would go to the extreme lengths we did. He would be more of a consumer rather than a producer.

My response. Deny, deny, deny.

The Coldstream pest was a lightweight. He was tenacious and was keen to learn the ropes. Over time, through our occasional chats, I noticed he had started developing a little bit more knowledge around computers, infrastructure, and how it all worked.

From the tone of his chats and the eagerness of his engagement, I was convinced he was a lonely boy. I imagined him being stuck out on a farm somewhere with no nearby neighbors. From memory, his parents were farmers who had a large property consisting of many acres on the outskirts of Melbourne.

During another of our chats, the Pest explained to me that the recent phone bill had come in, and his parents hit the roof. Apparently, he racked up over a thousand dollars' worth of phone calls, which was a very large sum of money at the time.

His parents had had enough of him, as he would constantly tie up the phone line so he could access the BBSs.

To be fair, we all did the same thing. My home phone line was tied up from between 9:00pm and 5:00am most nights.

I remember one of my uncles coming over and saying he had received a call from Greece complaining that they could never call our home phone number. They thought the phone had been disconnected because we didn't pay our phone bill!

My parents weren't too thrilled with the news, but I kept doing what I was doing.

Anyway, the Pest's parents decided to put a digital lock on the phone. They contacted Telecom Australia and implemented a 4-digit pin code to restrict the Pest from dialing out. The idea was simple: If he wanted to dial out, his parents would need to punch in a 4-digit code on the phone, which would unlock the phone and allow him to make a call.

Using some ingenuity and knowledge he learned from the boards, he managed to get a hold of a demon dialer program. The program was designed to dial every pin sequence from between 0-0-0-0 to 9-9-9-9. Numbers would automatically be punched in sequential combinations. When the code was identified, the program would capture those details. He would then be able to bypass his parents' locking system.

He explained all this to me over a chat, bragging about his triumph!

I thought to myself "Wow, okay, this kid's learning!". He showed some initiative to a problem he was keenly interested in solving, and used computers and telephony systems as a tool to achieve his ultimate objective.

Unfortunately for his parents, they kept changing the pin and he kept running his demon dialer program to break the code. In the end, they just gave up. They probably hoped he would grow out of his obsession.

I'm not sure what happened to the Coldstream Pest after that. With the incoming nuclear warhead heading our way, I lost touch with him. I never actually met him. He never attended any of the CBD catchups because he was obviously not at the same level.

The main issue for me was that the Pest was a bragger and a gossiper. I constantly thought to myself, "If this guy is revealing things to me in an open public chat forum, he surely cannot be trusted to enter the inner sanctum".

And I sure as hell wasn't going to vouch for him, as I knew that I would be thrown out the second he talked to the wrong people about what we were all up to.

History Repeats

File sharing and chatting were the main activities on the boards. Whilst there was no graphical user interface, the basic functionality was already in use in the mid-1980's, at least 6 years before the Internet was "invented".

These same features that drew us to the boards like moths to a light were hugely popular back then. They became insanely popular on the Internet.

The types of files being shared weren't just games or text files. We would share other things like music files. There were large music catalogs available which people could download and listen to on their computers. They were poorer audio quality and much larger file sizes than MP3 files. Remember, the MP3 audio format wasn't invented until the 1990's. These music files were in a raw format, good enough for you to be able to listen to snippets of songs at a time.

Video files were also available for download. However, these video files required very large storage. Many hobbyists didn't have a hard disk drive, as it was extremely expensive at the time. We had floppy disks which were restricted to 880kb storage space. Video durations were therefore quite small.

They were similar in duration to today's Snapchat or TikTok videos. The length would have been 20 to 30 seconds maximum. However, the thrill of downloading and watching a short video for a teenager was just as exhilarating and addictive back then as what it is today.

BBSs also had basic social media features. In addition to real-time chatting and messaging, users were able to pin items on a board for other users to view, such as an upcoming social gathering or event, or a particular item for sale. BBSs were local, so people posting something in the public section would usually receive responses from others in their area. It was a digital version of 'The Trading Post', without the advertising fees.

Buyers and sellers would negotiate a price for an item in the private chat rooms, then arrange a date and time to meet up for the exchange. Cash was the currency of choice, but barter was sometimes also an option.

The first commercially developed computers were called mainframes. These would typically take up entire floors, or even buildings, of large corporations or government departments. They required a dedicated team of computer operators to monitor and support them on a 24/7/365 roster.

Users would connect to these systems using terminals (also referred to as "dumb terminals"). Terminals consisted of a monochrome green or amber text-based screen, a keyboard, and a dedicated network connection through a modem back to the mainframe computer. There could be thousands of terminals all connected to a single mainframe computer, coast to coast.

All the applications and computing power were handled by the mainframe, and the terminals were simple input and output devices which didn't contain any memory or storage capabilities.

When personal computers were developed, they provided local memory, storage, and processing power. Effectively, they ushered in the dawn of the decentralized computer.

When personal computers became connected to the Internet, the transition back to a centralized model had begun.

Today, home computers accessing the Internet take advantage of centralized systems such as cloud computing, cloud storage and backups, and third-party platforms. The development of the Application Program Interface (or "API") allows services to run in the cloud, and the home computer to act as a terminal.

The topology of computers has therefore shifted from a centralized, to a decentralized, to now a hybrid.

Fundamentally, today's computer system topology has come full circle to the original mainframe designs of the 1940's, although much more powerful, far more scalable, and with a lot more functionality.

As we move into Web 3.0, we will start to see a more decentralized approach return, at the platform level.

Online bullying is not new. We faced the same challenges back in the day.

I remember one time when the Coldstream Pest was being attacked in the public forum by a group of users. They were making fun of him and where he lived, which was pretty mean in my mind. They had ganged up on him like a pack of hyenas circling a young, wounded gazelle, their jaws and sharp fangs snapping. They were relentless.

To his credit, the Pest defended Coldstream like a front-line soldier, giving as good as he got.

"Good for you dude!" I thought.

Often, I would swoop in and defend the Pest, beating off the hyenas with a few select blows. Due to my status on the boards, and the fact that I knew many of the users on the forums, they would retreat. They respected my position in the pecking order.

When you have a melting pot of adolescent teenagers all looking for social validation, often egos would clash.

In the real world, some of these hyenas may themselves be the victims of bullying due to their intellect, interests, or physical appearance. They knew first-hand the sting of the attacks, and how it affected their mental health and wellbeing.

But online, they would become bullies themselves. Suddenly, they had morphed into something they despised.

I think this has a lot to do with human nature. Deep down, at our core, most of us have a thirst for power. Some may even have a primal desire to gain pleasure at someone else's anguish.

The solution is simple. If you are in a position of power and respect, stand up against bullying. People will listen and respond.

Speak up, don't ignore.

A lot of the core underlying features of some of the most popular platforms on the Internet today were derived from the BBSs of the 1980's. Many today think these features are innovative and groundbreaking. The reality is that they are just supercharged versions of what we were doing back in the day.

Today's platforms have a lot more bells and whistles. They are all mobile-first, always-on and always connected. But fundamentally, at their basic level, they are the same.

Citibank Hack

Having reached the pinnacle of the global computer underground, I was recognized as one of the top people in the world when it came to having access to information.

I had come a long way from my time in the labs of Thomastown High School.

The funny thing was that people around me had no idea what I was doing. They saw the outputs of my efforts through my vast games catalog, without having a clue how I was getting access to these games.

I was the top pirate on the top bulletin boards in Australia, based on file sharing ratios. I was one of the top pirates when it came to file sharing ratios on other popular sites globally. I had access to every piece of software I could ever possibly want.

I was being invited to attend cracking sessions and pirate sessions in Germany, Holland, the UK, and the US. Of course, I couldn't attend. I was a 17-year-old high school kid with limited funds and no passport.

I had reached the goal I set myself four years earlier: Top of the global mountain. One of the 800-pound gorillas.

Nothing to me was more important than learning about computers. School was secondary, so my studies dropped off significantly. I really wasn't interested in high school and studying. I was more interested in learning about computers because I believed that was where my future would be, and that's where my life would take me.

I had this sense of power and privilege because I'd made it to the very top. I was mixing in the right circles with the right people at the right point in time. We were the absolute elite.

The global computer underground had arrived, and it was magnificent!

Suddenly, everything changed.

In 1988, Citibank happened.

CITIBANK HACK

For those of you who aren't aware, in 1988 Citibank announced publicly that they had been hacked. It made global mainstream news. In fact, it was broadcast all over the Australian television and newspapers. It made the pages of the German papers, UK papers, and went coast to coast across the US.

Initial reports stated that a team of US hackers had hacked into Citibank. They didn't spell out exactly what was done, only that it was a very severe hack. Later, it was announced that an estimated $500,000 had been stolen from Citibank. Personally, I suspect the number was a lot higher. At least, that was the rumor running wild through the community at the time.

Citibank, to their credit, stepped forward and admitted they had been hacked. They went on the front foot and announced it to the world. Immediately, the FBI got involved. They were tasked with figuring out what had happened.

In 1988, law enforcement was not what it is today. Most agencies were still using typewriters and handwritten internal memos to communicate. Computers were reserved for administrative and senior positions.

Importantly, law enforcement had limited tools at their disposal to investigate these types of crimes. They not only lacked the hardware and software needed to assist with their investigation, but they also lacked the knowledge and training, as well as the regulatory teeth to prosecute.

They were effectively bringing a knife to a gun fight.

The news coming out from America was that the Citibank hack had been perpetrated by a local US-based team. There was no way anyone else could have the capabilities, know-how, means, and desire to do it.

The global computer underground was buzzing with discussions around the Citibank hack and its implications. It was such a big news story.

We knew almost immediately that the hack was not a US-based team. We knew that, in fact, the hackers had come from Australia.

We also suspected that the hack had been done by the top hacking team in our ecosystem: The Realm.

These five guys from Melbourne, who made up The Realm, were the ones who did the hack. After all, only the best in the world could possibly pull this off.

There were a lot of signs that pointed to them. Firstly, immediately after the Citibank hack was announced, The Realm members went silent. Even more silent than before. They spoke to no-one, they chatted with no-one, they did not answer phone calls, they wouldn't answer their front door, and they stopped attending the private catchups.

We knew the FBI was involved. We knew they were looking inwards to a US-based team. We knew the hack was done from Australia.

Given all the above, the feeling amongst the inner sanctum was quite calm. We didn't think much of it, and believed things would just go on as normal.

After all, we were the masters of the universe. We were smarter, faster, younger, better equipped, and more driven than the authorities.

What could go wrong?

And so, we went on as we'd done in the past. We kept doing what we were doing. We kept having our regular catchups, we kept performing our nightly downloads, we kept doing everything that we were doing before, and we kept logging onto Zen. We kept chatting on Zen, and we kept posting on Pacific Island; thinking nothing much had changed.

Gradually, we started noticing new names appearing in the private section. Again, we didn't think anything of it, as we figured someone new had been invited to join our secret community.

Interestingly, these new users weren't turning up to the CBD catchups. They would be active on the boards, regularly chatting in the private forums, and asking questions about what we did.

It seemed that whenever I was on Zen this new person would almost always reach out to me. His name was Stuart Gill.

"Strange name…" I thought.

BBSs were a place where you could create your own persona different from your real life. It was a way to build out the character you always wanted to be but were too scared or embarrassed to reveal.

You would pick a name that was cool, that people would remember. A name which defines your online character.

Most people on the boards were teenagers, so the names were more in line with an adolescent audience. Killer Tomato, Masked Avenger, Interceptor, Blue Thunder, Rebel, Electron, Nom, Force. These were all cool-sounding names.

Stuart Gill was not a cool name. It was not a name you would choose to represent your alter-ego. In fact, it's the kind of name you would want to change. Teenagers wanted to have exciting online names. Stuart Gill was not exciting.

On top of that, people wouldn't choose other real-sounding names as their pseudonyms. Even guys like Ivan Trotsky had a commie angle to his online persona. No doubt this wasn't his real name, but it sure as hell was much better than Stuart Gill!

The only other person I knew who had a boring real-sounding pseudonym was the Zen and PI SysOp, Craig Bowen. But Craig was in his late-20's, 10 years older than most of his users. We all knew Craig Bowen wasn't his real name, but we also knew Craig from the Melbourne CBD meets and other social catchups.

Stuart Gill turned up out of nowhere. No-one had ever met him in our community.

As a community, we were always open and welcoming to new users who had been verified by our peers. We would chat with new people coming online.

Stuart Gill pestered me for probably three months. I'll give him credit, he was persistent. I chatted with him a few times, but something didn't feel right.

From the 'tone' of his typing, you could tell he wasn't one of us.

People type in certain ways and use a certain language range. For me, it was easy to guestimate the age of the person on the other end of the chat. Someone like the Coldstream Pest used short words, plenty of swearing, and not a lot of substance.

"Yep, he's a kid…" I would think.

Others like Blue Thunder were older, probably in their early 20's. His vocabulary range was broader, his sentence structure more refined, and his use of the English language more mature.

Masked Avenger was a classic juvenile delinquent. Every second word was a swear word, and he was constantly getting into arguments with others during chats in the public forum.

"Yep, short-fused teenager…" I thought.

When I met Avenger at the Melbourne CBD meets, he basically ticked every cliche box I had imagined. And to top it off, he wore the standard attire: flannel shirt, white T-Shirt, blue jeans, black boots.

I was very prominent on the pirate section of Zen. I didn't really go onto the hacking section often, although I used to go there every now and then to see the latest discussions about the Citibank hack. I was curious to see if there was any fresh news coming through.

It was January 1989. I had finished my Year 12 final exams and had just received my offer for university. I was 17 and enjoying the summer break. I remember being summoned by a nervous hacker through a private Zen chat the previous day. We would meet at the usual place: City Square in the Melbourne CBD.

The news was bad. Really bad!

The Australian Federal Police (or 'AFP') had infiltrated Zen and PI. The FBI had realized somehow that the Citibank hackers were based in Australia, and they had put pressure on the Australian Government to act. Somehow, somewhere, someone pushed a button and got the AFP involved. And this time they were taking it very seriously.

The AFP had been accessing Zen and Pacific Island for at least 3 months. They had infiltrated the private sections and had been building profiles on all the key players on the sites. Specifically focusing on The Realm.

I imagined a team of middle-aged, overweight, balding Feds sitting on desks and chairs in a dingy smokey room. All staring at a light-brown cork pin-up board illuminated by a flickering fluorescent light. Black and white headshots pinned in a hierarchical format, red string interconnecting the suspects, and brief hand-written notes pinned underneath each picture. Just like a Hollywood movie scene.

The news exploded like a mushroom cloud.

They had been using aliases to chat to people, gathering information to identify the people that were responsible for the Citibank hack.

Guys like Interceptor, Force, Nom, Electron, and Phoenix were Public Enemy Number One as far as the AFP were concerned, and they were obsessed with finding them and bringing them to justice. The FBI were riding shotgun, pulling every diplomatic lever to ensure the investigation powered ahead unabated.

Basically, every conversation that we'd had in the private sections of Zen and Pacific Island was now in the hands of the AFP. Every file that was shared, every ratio, every communication, every exchange of credit cards, of calling cards, of anything and everything.

They identified the members of The Realm who had hacked Citibank.

As a result, anyone who was associated with The Realm was being targeted.

I personally had private communications with members of The Realm. I didn't talk about Citibank; I didn't talk about the hack. I didn't talk about what they were up to.

Remember, the best hackers don't talk about what they do. They most certainly won't divulge anything on an electronic board.

They were usually very good at hiding their tracks. Most people would not be able to figure out what they were up to, and they made sure everything they had would be stored on remote servers, in rat holes, where they were the only ones who knew where they were and what they contained.

Well, the AFP came down hard. They started arresting members of The Realm. They shut down Zen and Pacific Island. The FBI, through their international counterparts, shut down other BBSs in the global community. They systematically shut down the whole ecosystem.

The music had abruptly stopped, and everyone was scrambling for life jackets.

As a 17-year-old about to enter university, it was panic stations for me. I felt I was now exposed. The inner sanctum trust layer had been breached. Everyone in the community was a potential rat. No-one could be trusted anymore.

I also worried that there could be possible jail time if I was somehow linked to the hack. Whilst I had absolutely nothing to do with it, maybe I knew too much. Maybe I asked the wrong question at the wrong time to the wrong person on a private chat.

Hollywood movies portray the FBI as an efficient well-oiled and well-funded machine, who always get their man.

Soon, there were talks of a new Computer Crimes Act being drafted in Australia.

I thought to myself: "Okay, I'm 17, I'm about to start university. The last thing I need in my life right now is to have these sorts of problems!"

So, I started getting rid of things. I got rid of all my software, login credentials, phone numbers, paperwork, even system reference manuals. Anything that linked me to Pacific Island, Zen, and every other global BBS was gone. Floppy disks, printouts, copies of passwords. The lot.

Everything disappeared. Torched. Gone.

I immediately cut all ties with the global computer underground. I never attended another CBD catchup, and stopped attending the computer swap meets which were held around town.

Cold turkey. No more. I was done.

I was resigned to the fact that, at any moment, someone would break down my door and say, "OK kid, you're coming with us".

Those were the most nervous few months of my life!

We were the most elite hackers and pirates on the planet. We had achieved everything, but we were left with nothing.

In the blink of an eye, it was all gone.

As teenagers, it hit many of us hard. Everything we had worked for was taken from us. By a system that was never our friend. All the hours spent online, contributing to the community, sharing laughs with friends we met from around the world. All of it, gone.

It was a hard pill to swallow.

It's not like we had any formal qualifications which we could take with us to our next job. Imagine someone from the community writing on their job application "I was once one of the best hackers in the world and hacked into several top secret and high-security sites including NASA, Citibank, and Department of Defense for fun. But you can trust me to look after your sensitive data and systems…"

It wouldn't take long before the Secret Service turned up on their doorstep to haul them away for 'questioning'.

We had lost faith in the world. We were nobodies again. We had sacrificed the most important years of our lives. For nothing. Zip. Zero.

As painful as it was, I had to face the harsh reality that it was all over.

We were evicted out of that life. Thrown to the curb like a used napkin. Left to meander aimlessly through the gutters of the real world.

Thankfully for me, there was a tiny sliver of hope.

I started focusing on my university studies, eventually met a girl (who later became my wife) and slowly put that part of my life behind me.

Looking back, it was fun while it lasted. I learned a lot. But ultimately, it was the right time to move on, even if it didn't feel that way at the time.

Life After Citibank

That one single act by The Realm to hack into Citibank resulted in the collapse of the entire global computer underground as we knew it, and rippled across other countries including the USA, Germany, and UK.

The whole ecosystem changed forever.

People were being arrested all over the world.

The new Computer Crimes Act was introduced in Australia, and similar new legislation was introduced in other countries around the world.

Law enforcement had woken up, got smarter, and set up dedicated departments specifically tasked with investigating cyber crime. What was once a joke in the FBI and AFP, suddenly became an important section of law enforcement.

Citibank was a big deal because it represented the establishment. Playing around with a research computer on some university campus was one thing, but going after the financial systems of one of the biggest and most prestigious banks in the world was something completely different. The integrity of the entire system had been drawn into question, and no politician or law enforcement organization would allow that to happen.

Implementing legislation was only part of the story. Companies wised up and decided to start spending more time and effort protecting their most valuable commodity: information. They hired experts to come in and check their systems to make sure they were safe. They paid millions of dollars for new equipment such as firewalls and network monitoring systems.

High-priced consultants were engaged to produce security guidelines to thwart any potential exposures. Password change rotation protocols were devised, along with other security related policies and procedures. Eventually, these evolved into industry standard operating procedures which were implemented by every major corporation on the planet.

Hackers, once the unwelcome squatters of computer systems, were being hired as "white hackers", tasked with identifying vulnerabilities in large corporate systems security. They were now highly paid corporate guys on the right side of the law.

All these activities led to the establishment of an entire new multi-billion-dollar industry: cyber security.

I like to compare these changes with what happened with another fledgling innovation way back in the early 1900s called the automobile.

Back then, cars were a curiosity for the rich and the powerful, but not really something for the masses. As such, governments and politicians let people tinker.

When the popularity grew exponentially, they sat up and took notice.

These vehicles needed structure and control. Pretty soon, road rules were devised which outlined how cars should be driven: who has right of way at intersections, which side of the road cars should be driven on, new street sign designs to indicate whether you stop or give way to others, etc. Entire government departments were then set up to license and certify drivers and vehicles so that they were safe for public use.

They had invented an entire regulatory framework and charged accordingly for the service.

The car came first, and regulation and constraints followed once mass adoption started.

In our own way, we had contributed to waking up the sleeping giant. Our exploits had served as the catalyst for change in the industry, which was desperately needed.

I firmly believe that many of the teenagers involved in the global computer underground went on to help build the Internet we have today.

Think about it. Do you think people who understood telephony, networks, programming, and infrastructure suddenly turned up in the early 1990's? They must have come from somewhere.

I often ponder: "Would the Internet exist in its current form if it wasn't for the 1980's global BBS ecosystem and the people involved in its rise and ultimate fall?"

Absolutely not.

BBSs were the precursor to the Internet in terms of technology, architecture, design, and above all, personnel.

I believe the Internet was built by these hackers, crackers, pirates, and phreaks.

During my 6 years between 1983 and 1989, I'd learned a hell of a lot while being involved in the community. After graduating high school, I went on to study computers at university.

With all the knowledge I had learned during my adolescent years in the global computer underground, I coasted through my degree. I already knew a lot of the stuff that was being taught at university. It wasn't challenging for me, so I created my own challenges.

I used to do other people's homework and major assignments for fun. I used to write code for my girlfriend and other friends studying with me. It became a running joke amongst the students that I should have been awarded five degrees, not the one I received. After all, I did five times more course work than anyone else!

Importantly, this was legal. No potential issues with anyone breaking down my door at 3:00am, grabbing me by the collar and hauling me out to an awaiting squad car.

After Zen and Pacific Island, I really needed that stimulation. That buzz I used to get back in the day when I was doing what I did.

I missed the community. I missed working with some of the brightest young minds in the world.

I was now in a university classroom with people who really didn't have the passion or desire for the field they were studying. Sure, students might specialize in certain areas, but no-one had the depth and breadth of knowledge we had back in the day.

Talking to my fellow university students, their idea of computer science was inserting a floppy disk in a drive and running the 'Install' program.

"Monkey work!", I used to think.

No-one cared how the disk worked, how data was stored on it, how it was retrieved by the computer, how the CPU interpreted the commands, how the memory stored and managed data, and how it was displayed on a monitor.

There was one clear upside to not being involved in the global computer underground. I found I had a lot more time to socialize. University parties, nightclubs, birthday parties, hanging out with friends at local street drag racing gatherings. The stuff most university students my age were doing at the time.

I discovered a new way to spend my time. When I was deeply entrenched in the underground community, my mind would be racing 24/7. Every moment not spent online was productivity lost. I had trained hard to become number one, and I wasn't going to let time slip away to allow others to take my place. I had leveled up to the top of the global mountain. I had everything I could ever want.

In the blink of an eye, it was all gone.

Adjusting to a 'normal' life was proving to be very difficult. I still had cravings. I wanted to jump back into the community to see what was happening. To find out what my friends were doing. To see if the coast was clear to come back. My addiction to that life was still there, and I yearned to somehow return. Normal life wasn't stimulating enough.

The allure to return was strong.

Competing thoughts raced through my mind.

"That part of your life is over. You were lucky to get out cleanly. Now, focus on your future..." I would say to myself.

I had successfully walked away, avoiding a potentially messy situation with the authorities, and was free. But nothing could replace the buzz of the global computer underground.

I needed to stay strong. I needed to hold tight. I needed to grow up.

Thankfully university filled a lot of the void I was experiencing. I finally had time to do something for *me*. Become a normal teenager. Fall in love. Party. Hang out with friends. Talk to people about anything and everything other than computers.

My life started filling up with new things. Important things. My mind was now absorbing other stimulants.

Love. Friendship. Brotherhood. Family. Knowledge.

Slowly but surely, I had weaned myself off my online addiction, and found life beyond the digital realm.

You're Hired

When I first entered university in 1989, the professors held a Town Hall meeting with the entire first year student body. They proceeded to gloat that the economy was booming, computer jobs were plentiful, and students would have no problem getting a job when we graduated. In fact, many of us wouldn't make it to graduation as we would most likely get picked up by a company to begin working before our final year.

The computing vocation, we were told, was in high demand.

I remember thinking to myself: three years of work and I'll be out in the workforce doing what I love. Legally. And getting paid to do it.

There was never any doubt in my mind that I would finish my degree. My parents had sacrificed a lot for me to get to this point. I was the first in our family to go to university and was determined to finish and get a degree. I knew my parents would appreciate it. It was for them, as much as for me.

The 1991 recession hit Australia hard.

Suddenly, all the jobs which were promised to us were non-existent. It's as though we had entered a cave in 1989 full of optimism and exited in 1991 to a job market abyss.

I was in my final year of study with three months left before graduating. Things were looking very bleak. Students all around me were disheartened with the countless graduate recruitment job rejections they were getting. No-one was hiring.

I contemplated getting back into the global computer underground. It was something I was really good at, and I had a strong global network of people; could I lean on them to get me started again?

In reality, it was not a career path. Going back to that life as an adult had too many risks, none greater than the new Computer Crimes Act which had been introduced to thwart guys like us from doing what we did.

My parents had instilled a lot of traditional values which ensured I stuck to the right side of the law. The last thing I wanted to do was get involved in the global computer underground and all the activities associated with credit card fraud and hacking. Whilst I knew I would be a few steps ahead of the rest, ultimately, I knew that one day it would all end in tears. So, that was not an option.

This frog would not jump into boiling water.

I happened to be in the computer lab one day working on a computer architecture project for my girlfriend (I had already completed and submitted mine, and another for one of my friends).

While in the lab, one of my friends was sitting opposite me working on a document using a word processor. I asked what he was up to, assuming he was also trying to finish the assignment.

Looking up from his computer screen, he told me he was finalizing his resume for a job opportunity. I asked which company he was applying to, thinking to myself that he's got little to no chance of getting anything as the market simply wasn't hiring.

He proceeded to tell me that Telecom Australia, Australia's largest telecommunications carrier had come to our campus in search of two recruits for their

Graduate Recruitment Program. Apparently, notices had been plastered all over the school and in the computer labs.

Not only had I missed the notices, but I had also missed the seminar Telecom held with prospective candidates. Moreover, resumes had to be submitted by 5:00pm that evening. It was a hard deadline.

I quickly shelved the computer architecture assignment and immediately began writing my resume. I asked my mate to share his so I could see the format he was using. He had spent time working with the university's career advisors and researched resume formats in the library, so had the right structure and layout.

I've always been very good at word processing. Importantly, I was a very fast typer, probably one of the fastest in the school. Years of practice typing away on my home computer had served me well.

I managed to knock up a resume in around 30 minutes. I included all the key points including my background, experience, and knowledge. I knew I had a deep understanding of computers, but needed to focus specifically on what they were looking for. The resume needed to be specifically targeted to the role they were looking to fill.

The position was for a Maintenance Programmer in their Training Services division. At the time, Telecom

Australia managed all staff training internally and would bill departments directly through "funny money"; a term used to suggest no real money was changing hands, but just budget transferred between internal departments.

Importantly, the programming language they were using was FoxPro, a DBase III derivative. I knew this language backwards. It was probably the second-easiest programing language you could learn, only slightly more difficult than BASIC. It included a relational database component which allowed you to build quite complex applications including records management, billing, tariffs and charges, and reporting.

The positions were for a fixed term of six months. But in my mind, it was a job. A very simple job, but a paying job by one of the biggest and most prestigious companies in Australia.

I submitted my resume that afternoon, just in time for the deadline, and was fortunate enough to get shortlisted for an interview. My friend who told me about the opportunity didn't get an interview; he wasn't too happy with me.

I went through the interview process and, because of my deep knowledge and understanding of computers and the programming languages they were using, I was offered the position.

I remember receiving the formal letter of offer and showing my parents. They were both so proud of me. It was one of the best days of my life.

Telecom only recruited two people across Australia that year, and I was one of them.

I couldn't have imagined that I would be working for a telecommunications company. The very company I was using to discharge my extracurricular activities in the global computer underground a few years earlier.

In truth, I probably knew more about Telecom than any other applicant. After all, I had access to reference manuals for a lot of their systems back in the day.

The irony wasn't lost on me.

In 1991, Australia along with the rest of the world suffered from a deep recession. I was very fortunate to be at the right place at the right time and be selected to join a graduate program. All my friends who had graduated in 1991 and 1992 took up to four years before they actually got a job in their chosen field. Many weren't so lucky, moving into other fields just to get a paying job. The industry lost a lot of great talent during those lean years.

My six-month duration turned into a 10-year stint with Telecom Australia (which was renamed to Telstra Corporation in 1992).

In 1997, I was offered a job working for Telstra Multimedia, in the Internet IT Products Division. Telstra Multimedia was established as a special purpose vehicle which held all internet-related services and assets, as well as the new Pay-Tv cable network. Branded 'Foxtel', it was a 50:50 joint venture between Telstra and Fox.

I had moved up the ranks quite quickly. By the time I was 26, I had been appointed the Network Services Manager for the entire Telstra Multimedia division, managing a team of 12 staff across Sydney and Melbourne.

While there, I oversaw the first high-speed broadband internet consumer installation in Wheelers Hill and was responsible for overseeing the entire broadband infrastructure backbone network for the country.

I was allocated an annual budget of $80 million which I would deploy to ensure the internet services of the entire country would continue to operate.

I had come full circle. It wasn't long ago that I was on the other side of the network, exploiting vulnerabilities for my own personal benefit. Now, I was responsible for ensuring the entire Australian internet backbone infrastructure stayed online, including the pipes which ran across the ocean to the USA and Asia.

As a 26-year-old, I had reached a corporate middle management position quickly, and was given all the perks of the position. I had my own car space, my own office, a personal bar fridge stocked with alcohol and other beverages, and a fully maintained company car.

I was one level below an executive and had been identified as a potential future leader who was destined for greater things within the Company. In fact, I was sent to specialized executive training programs and attended strategy workshops with other senior executives.

All of this was possible because I had built up the foundation and understanding of technology at a very young age, and I was able to apply that in my working life. I attributed my rapid success in the corporate world to what I'd learned in the previous ten years.

Knowing how software, hardware, networks, infrastructure, and telephony all came together, I was fortunate to be at the right inflection point to witness the mass adoption of the Internet.

It was a crazy time working for Telstra Multimedia. We got to play with the best technologies and innovations of the day.

For example, many wouldn't know that Telstra had a banking license which was a remnant of the old Postmaster General structure. In Australia, everything related to postal mail fell under the Postmaster General.

However, when this thing called the 'Telephone' was invented, the Government of the day decided to set up a new federal statutory body called 'Telecom Australia' and peeled all telecommunications into it.

Back then, as today, you could pay your phone bill at the local post office. You could also pay bills from other service providers and withdraw cash from the post office. Because Australia is such a vast country, most small towns don't have a local bank branch. These branches might be literally hundreds of miles away.

But most towns in Australia would have a post office.

The Government required the post office to act as a bank to service customers in remote communities. The Postmaster General therefore needed to have a banking license to accept payments and provide funds at its branches.

When Telecom Australia was set up, for some reason, the banking license was also vended into the new entity.

Fast forward to 1998, and Telstra was working in a joint venture with IBM to develop the world's first online payment network. Using Telstra's vast infrastructure and balance sheet, and IBMs know-how, the team spent two years building this revolutionary service. The plan was to launch it in Australia first, then work with IBM to deploy the technology globally to other carriers.

Fifty million dollars later, the project experienced constant delays and budget blowouts. To make matters worse, a new start-up had entered the market and was quickly capturing market share: PayPal.

The game was over. Telstra shelved the project shortly afterwards.

More Than a Game

Thinking about my accomplishments during the 1980's, I always return to the innocence of that high school computer lab in 1983. It was a moment in time which has stayed constant and clear in my mind.

It was the *exact* moment when I first fell in love with computers.

Hidden amongst the fog of adolescence was one of the most important lessons of my life. The thing which defined who I was and where I would take myself.

Snake Byte.

This basic yet highly addictive game defined my global computer underground journey.

Looking back, I realized that I was in fact the snake methodically weaving around the screen. The bugs and fruit the snake was eating were the knowledge and skills

I was accumulating during my exploits. The growing of the snake represented me growing as a person during each phase of my life. The completion of each stage by successfully maneuvering the snake through an opening represented my levelling up as I took the next step on the global computer underground ladder. The lab assistant calling time on my limited 20-minute session was the Citibank hack and the collapse of the ecosystem to which I had dedicated my entire adolescent life.

The game defined my online journey. It was right there in front of me 40 years ago and I didn't understand or appreciate its symbolism until now.

Ironically, I had chosen the name 'Cobra Strike' as my pirating name. I guess everything happens for a reason!

I often think about the double-life I led. The need to be someone else online. Not wanting to cross-contaminate the personas. Thinking back, it was exhausting wearing different hats.

The need to be someone different. To immerse yourself in a world where you are respected and successful, perhaps more so than the life you live in the real world.

I had jumped from an obscure 12-year-old to a rock star on the BBSs. As a child, you are always under the thumb of others. You are a servant to other masters. You often get ignored, teased, picked on, abused, and overlooked. But online, you are someone. You can be anyone you want.

The escapism of the virtual world is alluring, and highly addictive.

As a teenager, you want to grow up fast. You want to jump to adulthood as quickly as possible. You're never a 15-year-old. You're 'fifteen and a half' or 'almost sixteen'. You desire to leap forward, to get to a point where you can make your own decisions and do whatever you want. The online world provides this.

As a middle-aged man, I wish I could go back and tell my younger self to slow down. The half-year jumps aren't important. It's what you learn along the way that is far more valuable.

Teenagers desperately try to jump forward into adulthood, and adults desperately want to return to the safety blanket of being a teenager.

As a teenager, time moves slowly, and you are often in a rush to get somewhere. As an adult, time moves fast, and you want to slow it down.

The older you are, the faster time goes.

Being immersed in the community, I had inadvertently leap-frogged my juvenile years. I had become an adult at the age of 14, carrying the daily stress and pressures of competition and ambition for many years. Alone.

I couldn't discuss this with anyone in my family. My friends wouldn't have understood either. And the community was just like me. We were all of similar age, constantly working hard to keep up appearances, keep the reputation alive. To stay on top.

Talking about personal teenager things to your peers simply wasn't cool in the 1980's. It showed weakness. It suggested you weren't committed to the cause or lacked the testicular fortitude to be the best. Athletes face the same dilemma.

Today, opening up and talking about your problems is much more prevalent and encouraged.

Mental health has become a major problem in our fast-paced world. Teenagers now have support networks available to help them work through their personal issues. Friends are taught from a young age to step up and offer help when needed.

The 'Suck it Up, Princess' attitude is long gone.

I was fortunate to find a career path which I loved at such an early age and built out a life which I am proud to have lived. Computers have brought me great joy throughout my life and remain a key part of who I am today.

I have embarked on a few entrepreneurial ventures in my life. Some have been successful, some unsuccessful, and some are still a 'work-in-progress'.

In January 2005, as a first-time founder, I unveiled a new and innovative pre-production product at the Consumer Electronics Show (or 'CES') in Las Vegas. The 'InFusion' was an audio media player which allowed customers to listen to live streaming Internet radio, MP3 music, and FM radio, all from a single WiFi enabled device the size of a deck of cards.

I rented a small stand in the 'Start-up Alley' section of the show. The product was a prototype which hardly worked. It was so temperamental, it would often need to be restarted in order to fix the buggy software it was running.

To my surprise, the InFusion was selected as a top-three finalist in the Best of CES Awards in the highly coveted 'Portable Audio' category. It was selected ahead of the 'iPhone Photo' and 'Sony Walkman', which were both launched at the show that year. This prestigious awards program was run by the Consumer Electronics Association (or 'CEA"), the global industry body which ran CES.

At the awards ceremony, I was approached by a Creative Labs executive who congratulated me for my revolutionary product. He was fascinated with what I had managed to achieve with such a small budget and team. It reminded him of his early days in the industry when he first started out.

We briefly discussed the other products on display at the awards ceremony, as well as the latest products being launched at the show.

It was clear from our brief encounter that he was a technologist just like me. He loved to discuss innovation and trends in the industry. I really enjoyed talking tech with him.

When we finished our conversation, he politely thanked me for the chat, handed me his business card, and moved back to the Creative Labs awards stand.

When I returned home to Melbourne Australia, I looked up the Best of CES awards to review all the winners. I hadn't had a chance to do it when I was there.

To my surprise, I saw my Creative Labs friend in all the press. As I read his name, I froze.

He was Sim Wong Hoo, Founder of Creative Labs.

I had met the man who invented the Sound Blaster which revolutionized the global computer audio market. The same product I was admiring in the glossy magazines at Radioactive's house way back in 1987!

He was an idol in the computer hobbyist community.

I couldn't believe it! He had given me his business card and asked me to send him an email to follow up, and I didn't even read the card when he handed it to me. I had no idea who he was when we met.

I frantically searched for his business card to no avail. Somehow, during my long commute back to Melbourne from Las Vegas, I lost it.

"What an uber lamer!", I thought.

Losing my identity when the Citibank hack happened was a massive jolt to the system. Everything I had invested in and achieved was gone.

However, the next day the sun rose, and other opportunities presented themselves. I went on to finish university, get a great job, marry the love of my life, travel all over the world, and have two beautiful children.

Life wasn't over after Citibank, it was only the end of one phase.

Although one game may have ended, there is always an opportunity to start another twenty-minute lab session. To turn the egg timer over and go again, this time armed with the knowledge, skills, and confidence learned from previous experiences.

My journey demonstrates that anyone can start with nothing and reach their own pinnacle. It just takes hard work, patience, perseverance, self-confidence, a bit of luck, and finding something you're passionate about.

Most of us will start at the bottom, and it's up to us to work our way up to reach the heights and goals we set ourselves.

After all, why limit yourself to just one game on a single floppy disk.

Review Request

If you enjoyed reading my memoir, please consider leaving a review on Amazon.

I personally read every review, and they help new readers discover my book.

www.ingramcontent.com/pod-product-compliance
Lightning Source LLC
Chambersburg PA
CBHW032033290426
44110CB00012B/790